BLACK SUN

The Day of the Lord

BLACK SUN

The Day of the Lord

By

Randy Wills

Black Sun
The Day of the LORD

©2023 by Randy Wills

Distributed by

Ascension Multimedia ™
www.ascensionmultimedia.com
P.O. Box 498232
Cincinnati, OH 45249-8232

ISBN: 9798987828403

It may be most meaningful to the Reader to learn that the Author's interest in Biblical studies is more personal than it is academic. As a disciple of Jesus Christ for more than half a century, the Author's scholarship in the Scriptures has been (and still is being) developed under the tutelage of the Holy Spirit. Similarly, as the Apostle Paul, who wrote (in Galatians 1:11-12):

> "But I certify you, brethren, that the gospel which was preached of me is not after man. For I neither received it of man, neither was I taught it, but by the revelation of Jesus Christ."

The Author's understanding of the Word of God, has been carefully built stone-by-stone upon the Foundation laid in Scripture itself; and not derived uncritically from others' teachings.

After graduating *summa cum laude*, with a Bachelor of Science degree from Kettering College of Medical Arts, the Author studied Law and Public Policy at Regent University. He is a husband and father, and has served as a pastor and evangelist for many years.

Randy Wills can be contacted via email:
info@prophecyguide.com

Other books by the Author:
The Seven Seals in Prophecy and in History   (2014)

CONTENTS

Introduction

About a third of the Bible's text involves prophecy, the foretelling of future events. Among that great volume of revelation, more is written about the day of the LORD than concerns any other prophetic subject. And no wonder. The day of the LORD will be a time unlike any other. Ever.

Never before has there been a generation like this one. Over the course of the past century, the human population more than quadrupled; not counting the hundreds of millions who perished by wars and plagues during that same time. There are more souls alive on earth right now, than the total number of all who lived during the past four millennia! This generation is further distinguished from all others, by its stupendous technologies. Above all, this generation is exceptional for its pervasive selfishness and lawlessness—manifestations of a satanic nature.

By denying the knowledge of God and the existence of objective, universal Truth, this generation has become increasingly alienated from reality—though not from the consequences of that. Wicked philosophies and ideologies, conceived long ago and cultivated to recent maturity, now fill the world, and with deadly results. Lost in a wasteland of perpetual uncertainty and insecurity, human life is stripped of meaning and value. Never has the devil thus had more liberty to abuse mankind, by every imaginable temptation.

God has judged the world before. Of the countless millions that must have lived before the Flood, only eight souls did God preserve alive to carry on His eternal plans involving humanity and the earth. God is going to judge the world, again. The New Testament warns, in many ways, that the soon-coming judgment shall be no less severe than was the Flood in Noah's day. But this time, God is going to judge and punish not one but two worlds, together, i.e., the world of men, and that of evil angels and demons.

The modern-day Church has been lulled to sleep with speculative ideas about the so-called "Tribulation". But those ideas very little resemble what the Bible actually teaches about the soon-coming judgment of the world.

This book presents an unconventional view of what the Bible calls, "the day of the LORD". But unconventional does not mean un-Biblical. My aim has not been to be unconventional, but to discern the real truth of God's Word. Noah was, to say the least, unconventional. So, too, were Moses, and Elijah, as well as John the Baptist, and—most unconventional of all, Jesus. Each of those men passionately communicated true ideas that must have seemed unconventional to all but perhaps very few.

Let me first define what I understand, from Scripture, is meant by the phrase, *the day of the LORD*. In essence it entails a **transitional period** (a decade or less) between the end of this present (Church) Age, and the beginning of what is often called, the Millennial Kingdom Age. That brief *time* corresponds, approximately, with what by many is confusingly called, "the Tribulation". But that is about as far as any meaningful correspondence may be discovered between conventional ideas about the so-called "Tribulation", and what is presented in this book pertaining to the Biblical "day of the LORD".

The day of the LORD is that time when God will bring this present world Order utterly to an end by means of the severest judgments and, *in the process*, institute the Kingdom of Jesus Christ on earth. The day of the LORD, Jesus said, shall be the absolute worst time of trouble on earth. Ever. Past *and future*. Here is Jesus in his own words:

> "For then shall be great tribulation [trouble], such as was not since the beginning of the world to this time,

7

no, nor ever shall be. And except those days should be shortened, there should no flesh be saved...."

MATTHEW 24:21-22

Not only human life, but all life on earth will very nearly be wiped out, in the day of the LORD. Near total annihilation— of 8 billion souls! For comparison, about 60 million died in all of World War II. It staggers the mind to consider that God's wrath, during the day of the LORD, may result in a death toll surpassing World War II, by at least **125** *times or more!*

That dreadful reality, however, seems not to constrain the imagination of many who presume to tell a very different story about the day of the LORD than what is plainly described in Scripture. Yet it is impossible to rightly understand Bible prophecy, unless, one is willing to hear what God has to say *in his own words.*

Which brings us to the matter of hermeneutics (how to interpret Biblical text). Among all the books of the Bible, *the Revelation* is admittedly one of the most challenging to rightly interpret. Not only because it is rich in its use of symbolism. More important, *the Revelation* includes many references to real things, events, and/or circumstances that are so far removed from common experience that, even when described in plainest language, seem fantastic. Many individuals are therefore tempted to assume that such language must be symbolic— when, in fact, those passages actually describe real things, in literal terms. One must not approach *the Revelation* (or any other book in the Bible), insisting that the Word of God has to conform with common experience nor even with widely accepted claims of modern Science, so-called. Whether by 'spiritualizing' Scripture, instead of taking it for what it says and believing it; or, by mistaking literal, for symbolic, language: both of those errors are invariably misleading and confusing, to one's understanding of Scripture.

It should be self-evident that the *true text* of Scripture, the true Word of God, must be the object of all such studies.

Having dedicated many hundreds of hours, throughout the past nearly two decades of my life, to investigate the identity of the true text of Scripture: I am compelled to state my firm belief, that **the King James Bible is the true Word of God**—*in the English language.* The use of *any* other, English-language Bible 'version', cannot possibly lead to a right understanding of the truth of God. For, all such so-called versions are fraught with insidious deception.

Finally, it will also be helpful, for studying the material in the following pages, to understand the context for this present work. In my previous book **The Seven Seals in Prophecy and in History** (2014), I elaborated upon the true meaning of the Seven Seals described in *Revelation* 5:1 thru 8:1. In his heavenly vision, the Apostle John saw God on His throne, holding a certain book sealed with seven seals. Only the Lord Jesus Christ was worthy to open those seals and reveal the contents of that divine book (Rev. 5:1-5). Numerous theories have been suggested, concerning the identity of that book. Yet the answer is self-evident: That book's contents are known to us as *the Revelation*—the most complete description of events pertaining to the day of the LORD and the Second Coming of Christ, found in all of Scripture.

The Bible clearly shows that the opening of the last of those Seals, above alluded to, heralds the arrival of the day of the LORD. In **The Seven Seals in Prophecy and in History**, I discussed at length an extraordinary set of events uniquely associated with the Sixth Seal, which *set* of events I called, "the Cosmic Sign". The events comprising the Cosmic Sign are very specific:

> the (physical) heaven and earth are powerfully shaken

> the sun becomes "black"

> the moon becomes "as blood"

> the stars of heaven appear to be affected

9

Wherever those events appear together, throughout both the Old and New Testaments, they are always associated with the advent of the day of the LORD. The following passage identifies those events that, *collectively*, I refer to as **the Cosmic Sign**. Note the connection of those events with the Sixth Seal, as well as with the appearing of the *"great day of his wrath"*— that is, the day of the LORD:

> "And I beheld when he had opened the sixth seal, and, lo, there was a **great earthquake**; and the **sun became black** as sackcloth of hair, and the **moon became as blood**; and the **stars of heaven fell** unto the earth, even as a fig tree casteth her untimely figs, when she is shaken of a mighty wind. And the heaven departed as a scroll when it is rolled together; and every mountain and island were moved out of their places. And the kings of the earth, and the great men, and the rich men, and the chief captains, and the mighty men, and every bondman, and every free man, hid themselves in the dens and in the rocks of the mountains; and said to the mountains and rocks, Fall on us, and hide us from the face of him that sitteth on the throne, and from the wrath of the Lamb: for **the great day of his wrath** is come; and who shall be able to stand?"
>
> REVELATION 6:12-17

The following passage from the Old Testament reveals the same relationship between the Cosmic Sign and the arrival of the day of the LORD:

> "Behold, the **day of the LORD** cometh, cruel both with wrath and fierce anger, to lay the land desolate: and he shall destroy the sinners thereof out of it. For the **stars** of heaven and the constellations thereof shall not give their light: the **sun shall be darkened** in his going forth, and **the moon** shall not cause her light to shine. And I

will punish the world for their evil, and the wicked for their iniquity...."

ISAIAH 13:9-11

Though many like prophecies in the Bible could be referenced, yet, one more passage of Scripture will be given to show the connection between the Cosmic Sign and the day of the LORD:

"The **sun shall be turned into darkness**, and the **moon into blood**, **before** the great and the terrible **day of the LORD** come.... Multitudes, multitudes in the valley of decision: for the day of the LORD is near in the valley of decision. The **sun and the moon shall be darkened**, and the **stars shall withdraw their shining**. The LORD also shall roar out of Zion, and utter his voice from Jerusalem; and the **heavens and the earth shall shake**...."

JOEL 2:32; 3:14-16

Notwithstanding the preponderance of Scriptural evidence, showing, that the Seven Seals' judgments all *precede* the day of the LORD; unhappily, most Bible scholars and prophecy teachers insist that the judgments associated with the Seven Seals do not occur until *after* the day of the LORD has commenced. Such (mis)interpretation has led to the development and propagation of many false teachings, comprising much of what I have called, "mainstream prophecy teaching". One example of which, follows:

At least three decades ago, a number of well-known Bible prophecy teachers began to teach that a great Revival is going to occur during the Tribulation. One of the better known of those teachers, the late Dr. Jack VanImpe, said: *"The greatest Revival in the history of the world is going to happen during the Tribulation"*. He even went so far as to say that the reason for such Revival *during* the Tribulation, is because, *"God is going to*

11

give Laodicea a second chance" (from his 1997 video titled, "You Asked For It").

I can hardly imagine a more dangerous false doctrine. Yet, over the past couple of decades and more, that same teaching, amongst multitudes of professing Christians and churches, has effectually supplanted the truth. Such teaching is the result of misinterpreting the meaning of Scripture (in this case, concerning the Seven Seals, in *the Revelation*). Which the devil then uses for his nefarious purposes. It does greatly matter whether one rightly understands the Word of God.

In <u>The Seven Seals in Prophecy and in History</u>, I described the successive events and circumstances in recent history (since the mid-1800's until now) that correspond with the opening of the first five, of the Seven Seals. Not only have many judgments, pertaining to the Seven Seals, already been in effect throughout the past century and more. As a matter of fact, those horrors described in the Fifth Seal actually occurred, and in unprecedented measure, throughout at least the past half-century and more. More Christians were martyred in the twentieth-century, than were killed in the previous nineteen centuries *combined*. Since the turn of this present century, Christian martyrdom has continued—but at a rate *more than double* that of the twentieth-century. We are now about to witness those events belonging to the Sixth and Seventh Seals. Which is to say:

The day of the LORD truly is ready to appear.

It is going to be unlike anything you have probably ever been told or even imagined.

12

The Cosmic Sign

"And I beheld when he had opened the sixth seal, and, lo, there was a great earthquake; and the sun became black as sackcloth of hair, and the moon became as blood; and the stars of heaven fell unto the earth, even as a fig tree casteth her untimely figs, when she is shaken of a mighty wind. And the heaven departed as a scroll when it is rolled together; and every mountain and island were moved out of their places. And the kings of the earth, and the great men, and the rich men, and the chief captains, and the mighty men, and every bondman, and every free man, hid themselves in the dens and in the rocks of the mountains; and said to the mountains and rocks, Fall on us, and hide us from the face of him that sitteth on the throne, and from the wrath of the Lamb: for the great day of his wrath is come; and who shall be able to stand?"

REVELATION 6:12-17

The following is an imaginary scenario intended to portray, as realistically as may be possible, the unfolding of *real* events succinctly identified, but not explained, in the above-quoted Scripture. It is told as if from the perspective of one reporting those events *in real time*. The narrative, including the timeline, is based upon what I have learned, from Scripture as well as from extensive scientific research, pertaining to the cause of those extraordinary events. They are real phenomena.

Not only can the nature and relationships of those phenomenal events be rationally explained. But when, after many years of diligent searching, I perceived the *initial cause* that sets in motion the events described in the Sixth Seal, that insight then opened unto me a floodgate of understanding, regarding the *whole series* of events described as judgments,

14

which occur throughout the day of the LORD, following the opening of the Sixth Seal.

In an effort to avoid interrupting the flow of the narrative, in-depth discussion and explanation is reserved for a later chapter. My goal in the following is to reveal, inasmuch as may be possible, what actual horrors and destruction shall be experienced by those upon whom the judgments of God will be poured out when the day of the LORD begins. The narrative includes relevant passages of Scripture, at appropriate places, to show the correlation between the unfolding scenario and the truth of Scripture.

Tuesday, 8:50 a.m. (initiating event: time mark, 0:00)
Suddenly, the sun seems to have brightened—a lot! Though there are no clouds in the sky that would have blocked its light before....

11:50 a.m. (+3:00 hours 'post-event')
The sun is definitely brighter. Over the past few hours, it also seems to have gotten—bigger(?) Its brightness is unusually intense.

2:30 p.m. (+5:40)
The sun definitely appears larger than normal. Its increased size makes it seem as though it is somehow approaching the earth, or vice versa. The air feels strangely hot. There seems to be a feeling of static electricity in the atmosphere....

In that part of the world where sunset is then occurring, it looks like the horizon has been set ablaze. At that same time, in regions where it is early nighttime, the moon is extraordinarily bright, the sky eerily luminous. Still further around the globe, it seems as if dawn is breaking, as it were, in the middle of night.

7:07 p.m. (+10:17)

All of a sudden, the intense sunlight that bathed the atmosphere dims and then totally disappears, within minutes! Before anyone realizes what is happening, the whole world is engulfed in what can only be described as "thick darkness":

> "[T]he day of the LORD cometh, for it is nigh at hand; a day of darkness and of gloominess, a day of clouds and of <u>thick darkness</u>.... [T]he people shall be much pained: <u>all faces shall gather blackness</u>."
>
> JOEL 2:1,2; 6

7:12 p.m. (+10:22)

What only minutes ago was an exceptionally bright, clear sky, is now blacker than deepest midnight. Only a diffuse, faintly glowing halo surrounds that spot where just minutes ago the sun was. But the sun cannot now be seen. It, too, seems to have been swallowed up by some cosmic darkness.

Widespread panic is causing the roads and highways to become jammed with wrecks and stopped traffic.

On the side of earth where it is then nighttime, the moon—which during the past few hours was intensely bright, suddenly disappears from view.

7:14 p.m. (+10:24)

The indescribable darkness—covering the entire earth, is now accentuated by the abrupt appearance of extraordinarily brilliant stars! So many of which, it seems, have never been seen before. Stars! at this time of day?!

7:22 p.m. (+10:32)

The unnerving silence that throughout the past few minutes gripped the wakened world in awe of the mindboggling spectacle still unfolding, is shattered by such shrieks and groans of fear as have never been portrayed in any horror movie. Panic is everywhere.

16

"The great day of the LORD is near, it is near, and hasteth greatly, even the voice of the day of the LORD: <u>the mighty man shall cry there bitterly</u>. That day is a day of wrath, a day of trouble and distress, a day of wasteness and desolation, a day of darkness and gloominess, a day of clouds and <u>thick darkness</u>...."

ZEPHANIAH 1:14-15

"[T]he heart melteth, and the knees smite together, and much pain is in all loins, and the faces of them all gather blackness."

NAHUM 2:10

There are some, few, who understand what is happening. Thousands of wealthy individuals have been preparing for years, for just such an event as is now occurring. They have built for themselves what are called, 'deep underground BUNKERS'. They realize, too, they have very few hours, at most, to get to those hiding places tunneled deep into the mountains. When—a few hours ago, they were alerted by their contacts within various governmental agencies, that a <u>**nova-type event**</u> **involving our sun** is in progress, they immediately fled to those underground bunkers.

"For the <u>day of the LORD</u> of hosts shall be upon every one that is proud and lofty, and upon every one that is lifted up; and he shall be brought low.... And **they shall go into the holes of the rocks, and into the caves of the earth**, for fear of the LORD, and for the glory of his majesty, when he ariseth to shake terribly the earth. In that day a man shall cast his idols of silver, and his idols of gold, which they made each one for himself to worship, to the **moles** [underground] and to the **bats** [caves]; to go into the clefts of the rocks, and into the tops of the ragged rocks, for fear of the LORD, and for the glory of his majesty, <u>when he ariseth to shake terribly the earth</u>."

ISAIAH 2:12; 19-21

17

Wednesday, 01:30 a.m. (+16:40)

Power grids are now failing. Electrical transformers are exploding, everywhere. Cell phone and Internet connections are down. Pandemonium is spreading. Now, what light did still remain, in the midst of such thick darkness, is gone. It is as black as being deep in the belly of a cave without light. Against the inky background of space, strong Auroral lights are seen at all latitudes north and south.

11:00 a.m. (+26:10)

A morning without light! There is no way to get any news about what is going on. Everything has come to a dead stop. Nothing, anywhere, seems to be moving.

1:37 p.m. (+28:47)

More than a full day has passed since the massive explosion on the sun occurred. Suddenly, a thunderous roar swells up from below the ground— all around the globe: *earthquake!*

A magnitude 10.0 earthquake is unheard of. That would be nearly *32-thousand times* more powerful than a 7.0 earthquake! Still, even a 10.0 would not result in the kind of earthquake and destruction that is described in the Bible, in connection with the Cosmic Sign—Sixth Seal—events now occurring:

> "Behold, the LORD maketh the **earth** empty, and maketh it waste, and **turneth it upside do**wn, and scattereth abroad the inhabitants thereof.... [F]or the windows from on high are open, and the foundations of the earth do shake. The earth is utterly broken down, the earth is clean dissolved, the earth is moved exceedingly. The **earth shall reel to and fro like a drunkard**, and shall be **removed like a cottage**; and the transgression thereof shall be heavy upon it; and it shall fall, and not rise again."
>
> ISAIAH 24:1; 18-20

> "And I beheld when he had opened the <u>sixth seal</u>, and, lo, there was a <u>great earthquake</u>; and the sun became black as sackcloth of hair, and the moon became as blood.... [A]nd ***every mountain and island*** **were moved out of their places**."
>
> REVELATION 6:12-14

2:30 p.m. (+29:40)

The violent shaking has continued without letting up for almost an hour! Now, what?! Overhead in the pitch-black sky, the stars—they are beginning to move! Not like the imperceptible motion observed during any ordinary night. But they are *visibly* moving! It appears as if the stars are...falling to earth!

> "And I beheld when he had opened the sixth seal, and, lo, there was a <u>great</u> earthquake; and the sun became black as sackcloth of hair, and the moon became as blood; And <u>the stars of heaven fell unto the earth</u>, even as a fig tree casteth her untimely figs, when she is shaken of a mighty wind. And the heaven departed as a scroll when it is rolled together...."
>
> REVELATION 6:12-14

> "[T]he sun shall be darkened, and the moon shall not give her light, and <u>the stars of heaven shall fall</u>, and the powers that are in heaven shall be shaken."
>
> MARK 13:24

3:20 p.m. (+30:30)

The most violent shaking has subsided, but very strong quaking continues. The stars are no longer visible. When or how they disappeared—in the midst of such chaos!—no one knows. Perfect darkness. No electricity. No lights. The temperature has fallen significantly around the world. There are fires—lots of them!—but even the light of those seems to be devoured by the darkness.

Devastation like this has never been seen before. In addition to the unspeakable destruction caused by the earthquakes, unimaginably huge tsunamis are reaching, in many places, one hundred miles and more inland.

The moon is visible, again—but it is now hovering in a location very far from where it was last seen just a few hours ago. It is now the only light in the sky. Strangely, it seems to be emitting light from itself: not a bright light, but a strong, deep, glow the color of blood.

Thus far, the events portrayed all pertain to the Sixth Seal (Cosmic Sign). In reality, of course, those then ongoing events shall proceed unto their full manifestation—which includes other kinds of events that are detailed, in order, in *the Revelation*, in connection with opening the Seventh Seal and the series of 'trumpet' judgments then revealed. There are still more, unspeakably destructive events next to come, as the direct and ongoing results of the events above depicted. The scenario continues:

Seventh Seal

> "And when he had opened the seventh seal, there was silence in heaven about the space of half an hour. And I saw the seven angels which stood before God; and to them were given seven trumpets."
>
> REVELATION 8:1-2

Though not seen by anyone on earth, yet throughout the course of the Sixth and Seventh Seal events, the Court of Heaven—including the Lord Jesus Christ as well as numbers of angels, have solemnly been carrying into effect God's ancient decrees involving End-time judgments. Christ himself was the One who opened each of the Seven Seals in succession. But now seven angels appear before the throne of God, to each of whom a trumpet was given. They stand, waiting, as yet another angel comes forward—

> "And another angel came and stood at the altar, having a golden censer; and there was given unto him much incense, that he should offer it with the prayers of all saints upon the golden altar which was before the throne. And the smoke of the incense, which came with the prayers of the saints, ascended up before God out of the angel's hand. And the angel took the censer, and filled it with <u>fire</u> of the altar, and cast it into the earth: and there were voices, and thunderings, and lightnings, and an <u>earthquake</u>."
>
> REVELATION 8:3-5

5:56 p.m. (+33:06)

Another strong earthquake! But, what's this?! Fire—literal fire, falling—or, rather, *shooting* down from the sky! It's coming down like jets of rain! And the sky is beginning to glow orange-red! Fire is starting to spread everywhere on the ground! Houses—or, the debris of those; trees…everything that can burn is burning….

Thursday, 12:02 a.m. (+39:12)

The showers of fiery hail are now increasing rapidly in volume and intensity, like the sudden outburst of rain in a fast-moving thunderstorm. The whole atmosphere radiates with blinding light. Suddenly, fire is no longer coming in piercing streams from the sky…but the atmosphere is being ripped apart by great sheets of flaming plasma, miles wide, and traveling at supersonic speed! The earth shakes, again and again, as successive waves, of burning rocks and liquid fire, slam into the earth! Hurricane-force winds hurl flaming debris including rocks and trees and parts of buildings, like so much exploding shrapnel! The sound is beyond deafening; the sight, beyond description. There is little hope of survival, for anyone in this part of the world.

"And the seven angels which had the seven trumpets prepared themselves to sound. The first angel sounded, and there followed hail and fire mingled with blood, and they were cast upon the earth: and **the third part** of trees was burnt up, and all green grass was burnt up."

REVELATION 8:6-7

The expression, "the *third part* of trees was burnt up", does not mean that only one out of every three trees upon which the deluge of fire fell, was destroyed. Rather, it means that fully a **third of all the trees** *on the planet* were destroyed in the resulting firestorms. In all likelihood, an entire continent (or two) has been destroyed. That is to say, practically everyone living in that part of the world that was then facing toward the incoming blast from the sun, perished.

5:14 a.m. (+44:24)

"And the second angel sounded, and as it were a great mountain burning with fire was cast into the sea...."

REVELATION 8:8

The remaining material that was blown out from the body of the sun—or torn away from one of the inner planets, a couple of days ago, is about to arrive next. Because it consists of denser material, it is approaching at somewhat slower, though still ultrasonic, speed. It is massive. Scripture describes its appearance: "**as it were a GREAT mountain burning with fire**" (Revelation 8:8). A *great* mountain-sized object, measuring perhaps a mile (or much more) in every direction; a gigantic mass of white-hot fire—exploding and thrusting aside earth's atmosphere as it shreds earth's curtain like a high-velocity missle—

"And the heaven departed as a scroll when it is rolled together...."

REVELATION 6:14

22

As it happens, the earth is still turning, in a manner:

> "The earth shall reel to and fro like a drunkard, and shall be removed like a cottage..."
>
> ISAIAH 24:20

The scorched continents that, throughout the past few hours, were engulfed by the infernal blast, have turned out of the way of the incoming mountain of fire—which now strikes mid-ocean, with the equivalent to perhaps hundreds of millions of *megatons* of TNT, exploding all at once:

> "[A]nd the third part of the sea became blood; and the third part of the creatures which were in the sea, and had life, died; and the third part of the ships were destroyed."
>
> REVELATION 8:8-9

This is destruction on an incomprehensible scale. *Everything* in that vast body of water where that fiery mountain-sized object impacts, is destroyed, both above and below the surface. Not "a third" of that which is in that one ocean—but, with respect to the entirety of earth's oceanic covering, the destruction shall entail fully "a third" of the world's sea-life, and a third of all ships. All of which is quickly followed, in turn, by yet more, mind-numbing destruction...as gigantic tsunamis, hundreds (or perhaps thousands) of feet high, submerge cities along all the coastal regions of the entire (Pacific?) Ocean.

The foregoing characterization of events is not intended to be a dramatic exaggeration but, rather, a thoughtful elaboration of real events that are matter-of-factly foretold in Scripture. Within the first several days after the initial solar nova-type event, *hundreds of millions* of souls will perish. By comparison, the total number of casualties (military and civilian) in all of World War II, was about **60 million**. The solar nova and its aftermath—which is but *the beginning of God's judgments* in the soon coming day of the LORD, is likely to destroy—within the space of only a few days—perhaps fifteen times or more than died in all of WWII.

Even so, such a number (900 million souls) represents only about ten-percent or so of the current global population. Is there *really* going to be such mass destruction, during the day of the LORD? Yes, and so much more than that! In reality the destruction of human (and all other) life, which is going to occur during the day of the LORD, shall come very near to **TOTAL** annihilation. Jesus said: *"[E]xcept those days should be shortened, there should __no flesh__ be saved"* (Matthew 24:22). Jesus does not lie nor exaggerate, but he speaks perfectly the truth. When God arises to judge the world, in the day of the LORD, comparatively few will be (allowed) to survive.

God is not going to *completely* destroy the human race. Nevertheless, Scripture does indicate that the scope and scale of destruction shall resemble the judgment of the Old World, in which all but eight souls perished in the Flood.

> "But the day of the Lord will come as a thief in the night; in the which the heavens shall pass away with a great noise, and the elements shall melt with fervent heat, the earth also and the works that are therein shall be burned up. Seeing then that all these things shall be dissolved, what manner of persons ought ye to be in all holy conversation and godliness, looking for and hasting unto the coming of the day of God, wherein the heavens being on fire shall be dissolved, and the elements shall melt with fervent heat?"
>
> 2 PETER 3:10-12

All this, is just the beginning of the day of the LORD.

Black Sun

"And I beheld when he had opened the sixth seal, and, lo, there was a great earthquake; and the sun became black as sackcloth of hair, and the moon became as blood."

REVELATION 6:12

Ever since I first published **The Seven Seals in Prophecy and in History** (2014), I had an intense desire to understand the mystery of what I called, in that book, "the Cosmic Sign" (as explained in the Introduction). I knew the Cosmic Sign held the key to understanding numerous important prophecies related to the appearing of Christ for his Church, as well as concerning the arrival of the day of the LORD. Though I could not make any sense of the *physical processes* involving the Cosmic Sign events. I was most perplexed by the enigma of the 'black sun'. Then, too, there are stars falling to earth? And the moon, having the appearance of blood? What could it possibly mean? Are those descriptions of literal events? Or do those passages convey mystical meaning in symbolic language? The answer matters greatly to a right understanding of Bible prophecy.

I strongly suspected that the most significant revelation was to be found, not so much by considering those events *individually*. Rather, I believed the most important knowledge was hidden in some way that appears only when those events are viewed as being intimately *connected*, in the context of some other, *unnamed* event.

To my utter amazement, I discovered that the riddle of the black sun—which, for years had been one of the main things that hindered me from opening the treasure-box of revelation, so to speak, was the very key that ultimately enabled me to unlock that same box—and identify that unnamed event connecting the diverse events of the Cosmic Sign. The mystery,

26

involving the *physical phenomena* described in the Cosmic Sign, is both concealed and revealed by one and the same image; a key that both opens and shuts, as it were. That key is the black sun. Who could have guessed? that a black sun is a cardinal indicator associated with one of the most *luminous* and potentially *destructive* kinds of events in the cosmos: a **solar nova**.

Misleading theories related to a 'black sun'

One theory that was popular during the past few years, proposed that the Bible's imagery, involving a black sun as well as a blood-colored moon, could be explained in terms of solar and lunar eclipses, respectively. Advocates of that theory claimed to have identified a pattern involving a number of eclipses that, from time to time, occurred in close connection with certain Jewish Feasts described in Scripture. It was furthermore claimed that such pattern had occurred several times in recent history. But when none could identify any Biblical evidence, relating any pattern of eclipses to prophetic events still unfulfilled, that theory faded into obscurity, due to its ambiguity. Besides, though solar and lunar eclipses may be taken by some as omens, yet, eclipses are powerless to do physical harm. Not so, a black sun.

Others theorize that the appearance of a black sun and blood-colored moon could be caused by quantities of smoke or other pollutants blocking the light of the sun and discoloring the light of the moon. But that theory offers nothing to explain the incomparable earth-shaking and destruction described in connection with the Sixth Seal.

There are still other, less convincing theories than those just described. None of which, however, are at all compelling.

Recent Scientific discoveries

Fifteen years or so ago, I discovered the writings of Immanuel Velikovsky. Whose writings and theories, involving Cosmic Catastrophism, led me on a quest that soon introduced me to more recent work of numerous scientists and other scholars, who, throughout the past few decades, have been laying the foundation of a new Cosmology (study of the physical Universe), embodied in what is called, the **Electric Universe** (EU) **Model**. That new theoretical model has proved useful for developing many reliable predictions, as well as for explaining many natural phenomena that the so-called Standard Model of Cosmology is incapable to explain.

The Standard Model is based upon the notion that gravity is the primary 'force' that, following the (supposed) Big Bang, caused random clouds of space dust to come together to form stars and planets and then solar systems and galaxies. The Standard Model proposes that stars basically function as nuclear fusion reactors fueled by vast quantities of hydrogen; and, when that fuel is used up, the outer remnants of such fuel-depleted stars eventually collapse inward upon their core. The Standard Model imagines that such collapse in turn causes a reactionary *ex*plosion that destroys (or radically transforms) the star, in an event called a **supernova**. Thus, the Standard Model essentially allows for one, *and only one*, kind of explosion involving a star. Yet it is obvious that stars exhibit many, and sometimes gigantic, explosions *on their surface*. Which (along with many other kinds of cosmic phenomena) cannot reasonably be explained by the Standard Model based, as it is, on gravity.

The Electric Universe Model, on the other hand, postulates explanations related to structures and processes whereby stars have the ability to **exhibit a wide range of explosive events**—in size ranging from (least to most powerful): localized solar flares (which are vastly larger than earth); to, still larger explosions involving the entire star, i.e., micro-novas and even more powerful novas; and, finally, a supernova that can destroy a star.

The EU Model posits that stars are charged bodies that interact with cosmic-sized networks of electromagnetic energy in space—enormous rivers of electric current (so-called Birkeland Currents), some extending many thousands of light-years in length. According to the EU Model, stars can and do act in ways not unlike components in electronic circuits (similar to LED-type lights). Stars can vary greatly, over time, with respect to their electric potential—because, stars are integral parts of *dynamic* electrical systems. Unlike the Standard Model, the EU Model explains how that stars have the ability to do all sorts of things—besides producing only *steady-state* illumination or self-destruct in a supernova event.

Astronomers and Astrophysicists now have access to technologies that were not available until very recently. By the aid of which, contemporary researchers have lately made many new discoveries in fields related to and including Astronomy. Recently, for one important and relevant example, a number of stars have been identified which apparently display the hitherto unknown ability (of stars) to **disappear** and, later, **reappear**. A number of stars have even been *observed* to undergo nova-type explosions, *repeatedly*: a phenomenon that has been named, "recurrent nova".

Stars have recently been *observed* to go dark and, later, shine again.

Still-ongoing development of the EU Model, has relied heavily upon many discoveries made in the field of Plasma Physics, during the past half-century and more. Plasma Physics attributes the luminous (light-producing) quality of stars, in part to a certain condition of their outer 'shell' comprised of plasma. At sufficiently high energy levels, plasma emits photons of visible light: a state or condition called, "**glow mode**". At lower energy levels, plasma—though still consisting as such, does not emit visible light: a state or condition called, "**dark mode**".

Our sun is a star. It can go 'dark' and then turn back 'on', again—depending upon the energy level of the sun, within the context of the cosmic electro-dynamic system of which it is part. In fact, it can be proved that —

Our sun has gone completely dark before—at least twice within the space of *recorded history*.

And **that proof is in the Bible.** Those instances in Scripture, are also associated with other phenomenal occurrences—such as great earthquakes and fiery hail.

Instances of a black sun, in the Bible

The first time a black sun is explicitly described in Scripture, occurs in the story of the Exodus, where it is recorded that there was "thick darkness" in "all the land of Egypt" for three days. That event was preceded by a plague of "fire mingled with hail" that was so severe it did not merely burn but it "brake *every* tree of the field". The text references appear, in order, below:

> "And Moses stretched forth his rod toward heaven: and the LORD sent <u>thunder and hail, and the fire ran along upon the ground</u>.... So there was <u>hail, and fire mingled with the hail, ***very grievous***</u>, *such as there was none like it in all the land of Egypt since it became a nation.* And the hail smote throughout all the land of Egypt all that was in the field, both man and beast; and the hail smote every herb of the field, and **brake every tree** of the field."
>
> EXODUS 9:22-25

> "And the LORD said unto Moses, Stretch out thine hand toward heaven, that there may be darkness over the land of Egypt, even <u>darkness which may be felt</u>. And Moses stretched forth his hand toward heaven; and there was a <u>thick darkness in all the land of Egypt three</u>

days: **They saw not one another**, neither rose any from his place for three days...."

EXODUS 10:21-23

The effects attributed to each of the above-described plagues were extremely severe. The fiery hail—which, Pharoah said, was attended with "mighty thunderings" (Exodus 9:28), beat down virtually every tree in the land of Egypt, and killed many people and animals. The darkness was so thick the Egyptians could not even see one another!

It has been suggested by some—who want to discredit the Bible, that the darkness in Egypt at that time might have been caused by a volcanic explosion. But there are no volcanoes in or around Egypt. And no volcanic eruption has ever been known to produce darkness so "thick" that people literally could not see one another, for three days.

The scenario presented in the previous chapter ought not then seem unbelievable, in view of the severity of the plagues in the Exodus. (Not to mention any comparison with events in the Flood.) The events that happened in the Exodus; and, those events associated with the Sixth and Seventh Seals: appear too similar and in the same manner extraordinary, to be unrelated.

The second recorded instance in Scripture, involving a black sun, should be familiar to every Christian. It happened at the time of Christ's Crucifixion. Following, is the Bible's description of those events:

"And it was about the sixth hour, and there was a **darkness over all the earth** until the ninth hour. And the sun was darkened, and the veil of the temple was rent in the midst."

LUKE 23:44-45

"Jesus, when he had cried again with a loud voice, yielded up the ghost. And, behold, the veil of the

31

temple was rent in twain from the top to the bottom; and **the earth did quake, and the rocks rent**...".

Beginning about high noon, the sun was darkened for at least three hours. That darkness—which was not only in Jerusalem but it was *"over all the earth"*, could not have been the result of cloudy skies or a solar eclipse. At the same time, the veil of the Temple was torn in two from top to bottom, and rocks were "rent" (violently shattered) by a strong earthquake. It would have required great force to rip apart that massive veil, in such a manner as it was torn.

The foregoing case studies, in Scripture, prove that our sun is not only capable of blacking out and then reverting, again, to a light-producing condition. It has in fact happened at least twice within the past four-thousand years. It is possible, too, that a very large solar nova might have contributed to the global catastrophe of Noah's Flood.

By those same proofs, above discussed, the Bible also reveals that **a black sun is associated with events identical in kind to those appearing in the Sixth and Seventh Seals**, i.e., earthquakes and fiery hail. Those Biblical case studies, furthermore, clearly prove that our sun has, *in historical time*, manifested nova-type events—*of varied intensity and severity*.

A large-scale solar nova appears to be the best explanation for the entire set—*and sequence*—of events described in the Sixth and Seventh Seals.

The principle of 'Occam's Razor', applied to the solution here proposed, cuts deftly to the heart of the millennia-old enigma of the black sun.

Moon—as blood

The ghastly lunar spectacle described in the Sixth Seal is something altogether different from the so-called "blood moon" phenomenon sometimes caused by lunar eclipses. Such eclipses—if they ever did cause passing distress to the minds of some in primitive cultures; yet, those eclipses, which have occurred numerous times during the past few years, scarcely attract attention in the twenty-first century. But when the moon "becomes as blood", at the beginning of the day of the LORD, it is going to grip with fear the minds of all who will then see it.

As above discussed, there is *Biblical evidence* that solar novas have occurred before. Yet there is also *physical* evidence in support of the fact. It is perhaps not well known—though the information is easily obtained, namely, that multiple (NASA) missions to the moon were concerned with collecting and analyzing samples of vitrified materials (**vitrified: "converted into glass, or glass-like substance"**). On at least one such mission (which was video-taped, with audio), astronauts can be seen moving about on the lunar surface, and heard, saying, that there appeared to be "glass everywhere".

Vitrification is a process whereby rocks and minerals are converted into glass-like substances by melting those materials at temperatures well **above 1500** degrees Fahrenheit, followed by rapid cooling. But daytime temperatures on the moon peak at around **250** degrees Fahrenheit at its equator. The following quotation is the opening sentence in one NASA publication: *

> "The **ubiquity** [everywhere present] of **GLASS fragments** in **ALL** samples of lunar regolith material [ground cover] attests to **vitrification** as being a <u>major surface process on the Moon</u>." (emphasis added)

* Nash, Douglas B. and Conel, James E. (1973). *'Vitrification Darkening of Rock Powders: Implications for Optical Properties of the Lunar Surface'.* SAO/NASA Astrophysics Data System.

A solar nova is the most reasonable explanation, for "ubiquitous" glassy substances on the moon.

A solar nova—such as is capable to destroy one-third of all trees on earth, could reasonably be expected not only to produce **glassy materials on the moon's surface,** but also to produce **intense thermal radiation** that would cause the surface of the moon to glow: such that, from any vantage point on earth, would appear the color of blood.

'Off the charts' mega-earthquake

A thorough examination of this judgment, alone, could easily fill the pages of a large volume dealing with the subject. It would be a gross understatement to say that the earthquake associated with the Sixth Seal shall be an epochal event. Together with the solar nova, the ensuing mega-earthquake shall be like God's shutting of the door, on Noah's Ark. It will signal the sudden, violent end of this present Age.

If nothing more than the multiple passages of Scripture, related to this singularly important event and its consequences, were brought together in one place, skeptics would insist that such language and imagery cannot possibly be anything other than symbolic. But that doesn't change what God has decreed:

> "Behold, the LORD maketh the earth empty, and maketh it waste, and turneth it upside down....
>
> "...for the windows from on high are open, and the foundations of the earth do shake.
>
> "The earth is utterly broken down, the earth is clean dissolved, the earth is moved exceedingly.
>
> "The earth shall reel to and fro like a drunkard, and shall be removed like a cottage...."

ISAIAH CHAPTER 24

"And I beheld when he had opened the sixth seal, and, lo, there was a great earthquake....and every mountain and island were moved out of their places."

REVELATION 6:12-14

"Therefore I will shake the heavens, and the earth shall remove out of her place, in the wrath of the LORD of hosts,
and in the day of his fierce anger. And it shall be as the chased roe, and as a sheep that no man taketh up...."

ISAIAH 13:13-14

"The sun and the moon shall be darkened, and the stars shall withdraw their shining. The LORD also shall roar out of Zion, and utter his voice from Jerusalem; and the heavens and the earth shall shake...."

JOEL 3:15-16

God's voice is going to shake the heavens and the earth. God's voice is what will cause, somehow, the solar nova and the unimaginably great earthquake above described.   That earthquake (rather, succession of earthquakes)—*as characterized by Scripture*, are not only going to cause "every mountain and island" to be "moved out of their places". But the earth itself shall be "moved out of its place" (its orbit), evidently.

Phase I
  ➤ *"Every* mountain and island [shall be] moved out of their places."
  ➤ "The earth shall reel to and fro like a drunkard".
  ➤ "Behold, the LORD maketh the earth empty...and turneth it upside down."

Whether as a consequence of a magnetic pole-shift, or by literal shifting of the earth's crust: the entire surface of the earth is going to be suddenly and dramatically displaced. The earth itself shall "reel to and fro" like a staggering drunkard.

35

Phase II

➢ "The earth shall remove out of her place."

➢ "The earth...shall be removed like a cottage."

This appears strongly to imply that the earth is going to be removed out of its present orbit. The Bible declares that, at some time in the future (before or during the Millennial Kingdom Age), God is going to create "new heavens and a new earth" (Isaiah 65:17). Which most likely shall result from the earth being relocated to a different part of the galaxy: from which place, there would then appear, from earth's vantage point, a view of "new" heavens.

Phase III

➢ "And it [the earth] shall be as the chased roe, and as a sheep that no man taketh up...."

A "*chased* roe" suggests the idea of a small deer that frantically bolts and flees without any regard to its ordinary home-range, when chased by a determined predator. The imagery of "a sheep that *no man taketh up*", suggests a lone sheep ever wandering, away from the sheepfold and the shepherd. Those two metaphors depict the earth, first, as if fleeing away; thereafter, as a solitary, wandering body.

All of the above passages refer to the great earth-shaking that shall occur at the beginning of the day of the LORD. Taken together, those prophecies seem to suggest that, at minimum, the earth shall be violently shaken and moved out of its present orbital path around the sun. It may actually be the case that the earth shall remove completely from this solar system—a process which, in view of other events then still to come during the day of the LORD, would likely occur over the course of at least several years.

The foregoing Scriptural evidence is substantial. Yet, even more can be garnered from the Bible, pertaining to this subject. Notwithstanding such weight of evidence, however, those prophecies pertain to future realities that are so very far beyond the bounds of human experience—that few indeed have attempted to form an eschatological schema *involving those prophesied events.* But it is impossible to overlook or ignore such momentous revelation and still develop a reasonably accurate perspective regarding the true nature of the day of the LORD. If only the magnitude of that earthquake, and nothing else besides, were seriously considered (as herein): that one factor, alone, challenges every conventional idea about the 'Tribulation'.

The stars appear to fall

The Bible's characterization of that earthquake's immense power; as well as the Bible's explicit descriptions involving the dislocation of "every mountain and island", and the "reeling to and fro" motion of the whole planet: Such knowledge serves as the factual basis for imagining how that earth's chaotic movements might create the appearance of the stars falling. It shall also happen, at the time of that earthquake, that the entire earth shall be shrouded in "thick darkness". For some brief time, evidently, the stars shall be visible from every place on earth. Significant transient changes to earth's rotational velocity and/or direction might result in the appearance of stars rapidly moving in the sky, toward the horizon.

Another possible—and perhaps more likely—explanation involves fiery hail from the solar nova. Which in great numbers and large sizes shall appear, piercing and burning thru earth's atmosphere, from the highest altitudes—and causing ear-splitting explosions in the then-blackened skies. What else could that appear to be (in the prophet's Vision), but the stars falling to earth?

Solar nova is the 'best fit' solution

No other explanation serves so well as does a solar nova, to explain the characteristics and causal relationships of the diverse phenomena explicitly identified in connection with the events of the Sixth and Seventh Seals in Scripture. Recalling, of course, that the Seventh Seal entails still more, breathtaking judgments—which are discussed in the following chapter.

It will be seen that the solar nova's <u>explanatory power increases,</u> in order as we evaluate each kind of judgment that follows in the wake of that cosmic event.

Aftermath

"THE WICKED SHALL BE TURNED INTO HELL, AND ALL THE NATIONS THAT FORGET GOD."

PSALM 9:17

In stark contrast to God's merciful dealings with humanity in this present Church Age, God will pour out His wrath in waves of perfect judgment upon this present evil world until its destruction is complete, in the soon-coming day of the LORD. The spiritual darkness that shall overwhelm the earth when the Church is taken out of the world, at the Rapture, cannot now be grasped. The phrase, "hell on earth", will then in every sense be not a vulgar epithet but the inescapable reality.

Large portions of earth destroyed

As explained in the previous two chapters, the first of many *visible* judgments shall come in the form of a nova-type event involving our sun. The judgments that shall befall the world during the day of the LORD are not intended as mere warnings, but they are manifestations of God's unmitigated wrath.

The solar nova is not a warning shot: *it is a kill shot.*

The cosmic inferno that shall destroy fully one-third of all trees on the planet shall not be limited to some remote wilderness areas of western Asia, or northern Russia, or jungles of the Amazon. The nova-judgment is not against earth's trees. The description given in Scripture, pertaining to the destruction of "a third" of earth's trees, is intended to reveal the scope and severity of that judgment. Though the Biblical account is brief and to the point, yet the information is sufficient to allow many deductions and inferences to be made, concerning the nature of those events and their aftermath.

Impact zones

The initial, destructive events of the solar nova shall not at first affect the whole earth uniformly. It is possible to deduce, or at least infer with a high degree of confidence, the following:

1. what shall most likely be the primary impact zones;
2. the nature and extent of destruction in the early phase;
3. how those events shall consequently affect the rest of the world.

The Bible clearly identifies Israel as being at center-stage, as it were, amongst the larger theater of nations surrounding Israel, during the day of the LORD. The Bible also alludes in other ways to the preservation of certain national identities, unto the Millennial Kingdom era, represented by an apparently small number of souls from nations in and around the Middle East.

It thus appears that the destruction involving "a third" of the world's trees, by the solar nova judgment, does not pertain, *primarily*, to that part of the world where the Middle East shall be at the epicenter of later events. Ultimately, however, neither shall those parts of the world escape the solar nova's effects.

North and South America

The foregoing reasons lead to the conclusion that 'ground zero', of the solar nova's impact, appears most likely to be the Western Hemisphere (North and South America). It is perhaps significant that the Bible declares the solar nova shall destroy a third of the world's *trees*—and not a third of the world's *population*. North and South America contain a combined total of slightly more than a third (~ 40%) of the world's *forested* landmass, but less than fifteen percent (< 15%) of the world's population. Which may also explain why the

United States—prominent as it now is among the nations, is notable for its obscurity if not its complete absence (in prophetic Scripture), during the day of the LORD.

An Ocean and its borders devastated

The Bible reveals that the "great" mountain-sized object that shall impact one of earth's oceans (described in the Second Trumpet judgment), shall destroy every ship, as well as all sea life in that vast body of water. But that is not all. Cities all along the rim of that great ocean will be buried beneath tsunamis much taller than skyscrapers. If North and South America happen to be the primary impact zone of the solar nova, then that great mountain-sized object will probably explode in the Pacific Ocean. If that be so, then all the Pacific Islands, as well as all the coastal cities of Asia, Japan, Indonesia, and Australia—in addition to those of North and South America, shall be obliterated at that same time.

The Bible's descriptions of the scale of the solar nova's devastating effects, suggest that that event shall almost certainly cause very great disruption to other planets, especially, to those in the inner solar system. It is possible that the 'mountain-sized' impactor could be a fragment torn from Mercury or Venus—if it is not solid material from the sun itself.

Third and Fourth Trumpet judgments

"And the <u>third angel</u> sounded, and there fell a great star from heaven, burning as it were a lamp, and it **fell upon the third part of the rivers, and upon the fountains of waters**; And the name of the star is called Wormwood: and the **third part of the waters became wormwood** ["bitter"]; and **many** men died of the waters, because they were made bitter.

"And the <u>fourth angel</u> sounded, and the third part of the sun was smitten, and the third part of the moon, and the third part of the stars; so as **the third part of them was darkened, and the day shone not for a third part of it, and the night likewise.**"

The Third and Fourth Trumpet judgments are also a continuation of events associated with the solar nova. Fallout from the sun shall consist not only of an array of charged particles in ultrafast-moving plasma. But a veritable <u>toxic soup of elements</u> both from the sun as well as from the runoff of molten earth-borne materials, combined with the downpour of toxic rain and the fallout of airborne ash, shall pollute streams and rivers and aquifers throughout the world. The Third Trumpet thus signifies yet another widespread and deadly judgment, namely, the waters are made "bitter".

In the Fourth Trumpet judgment, the oft-repeated phrase, *"the third part"*, means the amount of light reaching the earth, from the sun, moon, and stars, shall be reduced by fully a third (**33%**). The continents-sized conflagration caused by the solar nova—together with an immeasurable volume of vaporized ocean water, shall produce a stupendously thick shroud of smoke and water vapor that will encircle the earth for many months or even years then to come.

Ice Age conditions

The effects just described shall combine to create ideal conditions for the rapid onset of a new Ice Age. The production of torrential rains and icy hailstorms, is an unexpected consequence of a colossal deluge of fiery material from the sun. The Apostle Peter prophesied that *"the elements shall melt with fervent heat"*. The nightmarish firestorms will create hurricane-force winds to feed the oxygen-hungry flames, which shall greatly intensify the heat and loft immense quantities of vaporized materials miles high into the upper atmosphere. Water vapor generated by the gargantuan explosion in the ocean shall contribute hugely to the indescribable volume of those gases and particulates. Earth-enshrouding clouds of super-heated gases, ash, and water vapor shall then cool and condense and congeal: thus, forming gigantic hailstones, with massive floods of ice and rain, of Biblical proportion:

> "And there fell upon men a **great hail out of heaven, every stone about the weight of a talent** [between 100-200 pounds each]: and men blasphemed God because of the plague of the hail; for the plague thereof was exceeding great."
>
> REVELATION 16:21

The combined and persistent effects of that dense, earth-encircling cloud cover, shall considerably block the sun's light from penetrating to warm the earth. Global temperatures will significantly and rapidly decrease—at the same time that unfathomable amounts of (toxic) icy rain and hail begin to pour down in torrents from above.

Extinction-level event

It is easy to lose sight of the fact that, in the context of the vast scale and scope of devastation hereto outlined, *only a few weeks or months at most* are suggested to have passed, from the initial solar nova event. In developing this analysis, care has been taken not to exaggerate or otherwise misrepresent the scale or processes of destruction; but, rather, to thoughtfully elucidate the true meaning and implications of the Biblical revelation. It is therefore remarkable that the Bible does not go so far as to portray the solar nova and its aftermath as being an 'extinction-level' event.

Or, does it?

Does the fact, that life on earth shall continue for several years after the solar nova, mean that human life could then go on indefinitely? If that were the case, then why did Jesus say that God must "shorten" those days—else, *"no flesh* would be saved"?

What does an extinction-level event actually look like? Who knows, really, how that any possible extinction-level event could proceed to waste the entire planet and its ecosystems? Must that occur more or less instantaneously? Or might it transpire over the course of a few years—as that is actually described in considerable detail, in Scripture?

There are other judgments yet to come, during the latter part of the day of the LORD, which exhibit still-worsening conditions caused by ongoing effects of the solar nova:

> "And I heard a great voice out of the temple saying to the seven angels, *Go your ways, and pour out the vials of the wrath of God upon the earth.* And the first went, and poured out his vial upon the earth; and there fell a noisome and grievous sore upon the men which had the mark of the beast, and upon them which worshipped his image.
>
> "And the second angel poured out his vial upon the sea; and it became as the blood of a dead man: and **every living soul died in the sea**. And the third angel poured out his vial upon the rivers and fountains of waters; and they became blood....
>
> "And the fourth angel poured out his vial upon the sun; and power was given unto him to scorch men with fire. And men were **scorched with great heat**, and blasphemed the name of God, which hath power over these plagues: and they *repented not* to give him glory.
>
> "And the fifth angel poured out his vial upon the seat of the beast; and his kingdom was full of darkness; and they **gnawed their tongues for pain**, and blasphemed the God of heaven because of their pains and their sores, and *repented not* of their deeds....

45

"And there fell upon men a great hail out of heaven, every stone about the weight of a talent [100-200 lbs.]: and _men blasphemed God_ because of the plague of the hail; for the plague thereof was exceeding great."

REVELATION 16:1-21

Obviously, the sun shall very soon revert to shining again, after the initial blackout. Then, within days of that initial event, sunlight shall be significantly blocked from reaching earth. A few years later, however, the light and heat from the sun will then beat down unbearably upon earth and its inhabitants; whether due to the ozone layer having been stripped away; or, due to the earth's nearer approach to the sun, as the result of earth having been "moved out of her place" (as earlier discussed).

The solar nova is, in fact, an "extinction-level" event.

—Nevertheless, God himself will eventually intervene, so as to preserve a sparse remnant, in order to prevent the total extinction of humankind.

Iron Age conditions

"[T]hey shall move out of their holes like worms of the earth...."

MICAH 7:17

In the days and weeks following the initial world-changing events, the survivors—especially those, elite, who had access to underground bunkers, will emerge from their afore-prepared hideouts—like _"worms moving out of their holes"_, not yet realizing the true condition of their plight. Prior to the solar nova, they were ready to rule the world. Now, everything has suddenly changed—far more than any then still alive could have anticipated.

All satellites then orbiting the earth, shall be destroyed by the solar nova.

Without those satellites, the worldwide Internet system shall be irrecoverably lost. The electrical distribution network ('grid') will catastrophically fail. Telecommunications will be non-existent. Manufacturing; energy and food production and distribution; transportation of all kinds; water purification and distribution: the most basic operations integral to sustain modern life shall cease, everywhere on earth (in developed nations).

The entire superstructure of the global technological Society will totally collapse.

Scripture reveals that potable water and food resources shall be scarce, because of the solar nova. Global conditions shall in very many ways preclude a return even to something like the pre-Industrial Age—when most people around the world could and did grow their own food and raise their own livestock. Those who live in so-called developed nations of the world shall be most severely affected by the complete loss of modern technologies.

Still, human life will struggle on for a little while to survive; though death in diverse forms shall already have begun to carry away many hundreds of millions of souls, as if by unstoppable rivers of destruction. Very soon after the solar nova, those who survive the initial onslaught will realize that the world they once knew, is gone. Forever.

Supernatural

"AND THE FIFTH ANGEL SOUNDED, AND I SAW A STAR FALL FROM HEAVEN UNTO THE EARTH: AND TO HIM WAS GIVEN THE KEY OF THE BOTTOMLESS PIT. AND HE OPENED THE BOTTOMLESS PIT...."

REVELATION 9:1-2

It is a serious mistake, one made by very many who, supposing the language and images of prophetic Scripture are too fantastic to be taken literally, they twist and distort the meaning of those words, in their efforts to derive from them some kind of 'spiritual' interpretation. In such persons' teachings, for example, the Bible's identification and description of a murderous "beast that ascendeth out of the bottomless pit"—somehow becomes: a <u>human being</u> who arises <u>amongst the world of men</u> to become *a* <u>charismatic world leader</u>! Who, then, could possibly connect that 'interpretive' description of the 'Antichrist', with the Bible's identification of one called, "the Destroyer", from the abyss? It is no wonder that Bible prophecy remains so mysterious, confusing, and controversial, to so many. Though it doesn't have to be.

Super-natural does not mean *unreal*

The day of the LORD is the climax of a dramatic, transitional period between two profoundly different Ages. Those two Ages—the present Church Age and the coming Millennial Kingdom Age, respectively, shall be more radically different from each other than, say, the current Space Age is different from the Stone Age.

To contemplate, beforehand, the many strange and inexplicable things that humans will experience during that transitional period, presents innate challenges to our ability to understand those unfamiliar subjects. Because, humanity is about to come face to face with realities, both physical and spiritual, that now seem impossible and unreal.

In seeking to understand that transitional period, it is helpful to start by accepting the end-point, foretold in Bible prophecy, as the coming reality. It is then easier to understand other, prophesied events that are nearer at hand, in light of that given, future reality.

Which of course implies that one must first have a fairly good understanding of what that end-point reality is going to be. In the case at hand, that future reality is the Millennial Kingdom Age. As God would have it, he has provided sufficient information in his Word, regarding the nature of that coming Age, so that we may acquire an essentially true perspective of that thousand-year epoch. If that coming Age should be characterized by any one word, that word is: *supernatural*.

New heavens and new earth

Let us then, as it were, journey backwards in time (in our minds), starting from the early days of the coming thousand-year reign of Jesus Christ on earth. By that time, the earth itself shall have been radically transformed to be the welcoming homeland of Christ's Millennial Kingdom. There is compelling evidence in Scripture, showing, that God intends to use the solar nova, together with other judgments during the day of the LORD, to accomplish his purpose and work in creating "new heavens and a new earth", for Christ's earthly Kingdom:

> "For, behold, I create **new heavens and a new earth**: and the former shall not be remembered, nor come into mind.... There shall be no more thence an infant of days, nor an old man that hath not filled his days: for the child shall die an hundred years old; but the sinner being an hundred years old shall be accursed. And they shall build houses, and inhabit them; and they shall

plant vineyards, and eat the fruit of them. They shall not build, and another inhabit; they shall not plant, and another eat: for **as the days of a tree** [longevity] are the days of my people, and mine elect shall long enjoy the work of their hands....

"For, behold, the LORD will come with fire, and with his chariots like a whirlwind, to render his anger with fury, and his rebuke with flames of fire. For by fire and by his sword will the LORD plead with all flesh: and the slain of the LORD shall be many....

"For as the new heavens and the new earth, which I will make, shall remain before me, saith the LORD, so shall your seed and your name remain. And it shall come to pass, that from one new moon to another, and from one sabbath to another, shall all flesh come to worship before me, saith the LORD. And they shall go forth, and **look upon the carcases** of the men that have transgressed against me: for their worm shall not die, neither shall their fire be quenched; and they shall be an abhorring unto all flesh."

ISAIAH 65:17-22; 66:15-16, 22-24

The above passages, excerpted in order from the last two chapters of Isaiah, clearly define the context wherein the "new heavens and new earth" shall appear. In preparation for the beginning of the Millennial Kingdom Age, Christ himself shall first appear, together with his holy angels and his then-glorified saints, to put down all rebellion on earth, and subdue not only all the nations on earth but also the whole dark empire of Satan's dominion. Myriad angels of God shall appear from heaven:

"...in flaming fire taking vengeance on them that know not God, and that obey not the gospel of our Lord Jesus Christ: who shall be punished with everlasting destruction from the presence of the Lord...."

2 THESSALONIANS 1:8-9

Christ will "render his anger with fury" and the "slain of the LORD shall be many", as Isaiah wrote. Thus, will Christ begin to manifest his power and authority over all the earth, thenceforth, forever. The "carcasses" of those slain, shall still remain—during the earliest days of the Millennial Kingdom Age—as the only reminder, apparently, of the world that formerly was.

In describing those same events and that same time period, the prophet Malachi wrote:

> "For, behold, the day cometh, that shall burn as an oven; and all the proud, yea, and all that do wickedly, shall be stubble: and the day that cometh shall burn them up, saith the LORD of hosts, that it shall leave them neither root nor branch. But unto you that fear my name shall the Sun of righteousness arise with healing in his wings; and ye shall go forth, and grow up as calves of the stall. <u>And ye shall tread down the wicked; for **they shall be ashes under the soles of your feet** in the day that I shall do this</u>, saith the LORD of hosts."
>
> MALACHI 4:1-3

The Apostle Peter also prophesied the appearance of "new heavens and a new earth", in connection with the day of the LORD:

> "But the day of the Lord will come as a thief in the night; in the which the heavens shall pass away with a great noise, and **the elements shall melt with fervent heat,** the earth also and the works that are therein shall be burned up. Seeing then that all these things shall be dissolved, what manner of persons ought ye to be in all holy conversation and godliness, looking for and hasting unto the coming of the day of God, wherein the heavens being on fire shall be dissolved, and the elements shall melt with fervent heat? Nevertheless

51

we, according to his promise, look for **new heavens** and a **new earth**, wherein dwelleth righteousness.

Earth's (atmospheric) heavens—"being on fire shall be dissolved", and the very elements of the earth shall "melt with fervent heat". All during the day of the LORD. Then, "new heavens and a new earth" shall become home to those, comparatively few mortals whom the Lord will spare their lives by "shorten[ing] those days" of unparalleled destruction (Matthew 24:22; Mark 13:20). Those same survivors—during the earliest days of the Millennial Kingdom Age, shall still be able to see (on the "new earth") the carcasses of those whom God did judge, in the day of the LORD.

The former earth shall not then be obliterated, or disappear, or be replaced by an altogether new earth. Nor shall the starry heavens of the Universe be altered. But the earth that now is, together with the solar system of which earth is now a part, shall suffer cataclysmic changes throughout the day of the LORD, such that earth's topography and climate will be greatly changed. Even earth's position in space, with respect to the present solar system, shall evidently be very different than that now is.

Reliable witnesses

Many images and ideas presented in prophetic Scripture are descriptions of real things—from the perspective of holy men of God who lived, and who saw in visions and wrote about those things, thousands of years ago. The Prophets and Apostles personally witnessed many supernatural things during their lifetime. The realm and operations, both of angels and of demons, were more than a little familiar to many writers of Scripture. It is important for us to know that the writer of *the Revelation*, as well as other writers of Scripture, wrote about things they knew quite a lot about, experientially.

The devil and his angels come down to earth
(Still working our way backwards in time, as it were...)

Those same angels of God that the Apostle John saw and interacted with, from time to time, are going to overthrow the Devil and his angels, and cast them down from out the heavenly realm into the earth. John wrote:

> "And **there was war in heaven**: Michael and his angels fought against the dragon; and the dragon fought and his angels, and prevailed not; neither was their place found any more in heaven. And the great dragon was cast out, that old serpent, called the Devil, and Satan, which deceiveth the whole world: **he was cast out into the earth**, and **his angels were cast out with him**."
>
> REVELATION 12:7-9

Which explains the awful warning given by a certain angel, who announced the fearful outcome of that particular judgment, during the day of the LORD:

> "Woe to the inhabiters of the earth and of the sea! for <u>the devil is come down unto you, having great wrath</u>, because he knoweth that he hath but a short time. And when the dragon saw that he was cast unto the earth, he persecuted...."
>
> REVELATION 12:12-13

The bottomless pit is opened
(As we approach nearer, going backwards in time, unto the earliest weeks and months of the day of the LORD...)

Prior to the Devil and his angels being cast down into the earth, the dark abyss—which the Bible calls, the "bottomless pit" shall first be opened: thus, unleashing unfathomable evil from that ancient prison beneath the earth. It is a place of unimaginable and ceaseless torment for those creatures who, in ages past, did make themselves to be God's enemies. The bottomless pit is going to be opened and its pain-maddened

53

captives loosed for a very little season on earth—at a time almost immediately following the solar nova event:

> "And the fifth angel sounded, and I saw a star fall from heaven unto the earth: and to him was given the key of the bottomless pit. And he opened the bottomless pit...."
>
> REVELATION 9:1-2

The prisoners of hell from below, and the rulers of Darkness from above...will all be brought together upon the earth, for judgment, during the day of the LORD.

All of hell from below, and from above, will come together on the earth, during the day of the LORD.

Even the demons fear to be confined in the bottomless pit. When Jesus cast out a "legion" of devils from a certain man, those wicked spirits implored Christ not to send them into the bottomless pit:

> "And Jesus asked him, saying, *What is thy name?* And he said, *Legion*: because many devils were entered into him. And **they** [the devils] **besought** [intreated] **him** [Jesus] that he would not command them to go out **into the deep** [Greek: ἄβυσσος "abyssos"—the bottomless pit]."
>
> LUKE 8:30-31

Torment from out of the pit

How dreadful is that place? How raging mad with cruel hatred must its prisoners be? God knows. And he has told us in his Word, what kinds of beings shall soon arise from out of the abyss to torment mankind, during the day of the LORD:

"And the fifth angel sounded, and I saw a star fall from heaven unto the earth: and to him was given the key of the bottomless pit. And **he opened the bottomless pit**; and there arose a smoke out of the pit, as the smoke of a great furnace; and the sun and the air were darkened by reason of the smoke of the pit. And <u>there came out of the smoke</u> **locusts** <u>upon the earth: and unto them was given power,</u> **as the scorpions of the earth have power**. And it was commanded them that they should not hurt the grass of the earth, neither any green thing, neither any tree; but only those men which have not the seal of God in their foreheads [the 144,000 are thus protected]. And to them it was given that they should not kill them, but that they should be **tormented** <u>five months</u>: and <u>their torment was as the torment of a</u> **scorpion**, <u>when he striketh a man.</u> <u>And in those days shall men seek death, and shall not find it; and shall desire to die, and death shall flee from them.</u>

"And the shapes of the locusts were like unto horses prepared unto battle; and on their heads were as it were crowns like gold, and their faces were as the faces of men. And they had hair as the hair of women, and their teeth were as the teeth of lions. And they had breastplates, as it were breastplates of iron; and the sound of their wings was as the sound of chariots of many horses running to battle. And they had tails like unto scorpions, and there were stings in their tails: and <u>their power was to hurt men five months. And</u> **<u>they had a king over them, which is the angel of the bottomless pit</u>**<u>, whose name in the Hebrew tongue is Abaddon, but in the Greek tongue hath his name</u> **Apollyon**."

REVELATION 9:1-11

Those are not the words or the wild imaginings of a science-fiction writer. They are true revelations of God. They are not meant to be 'spiritualized' and reduced to harmless nothingness.

They are literal descriptions of real events soon to come upon the earth.

First, the solar nova with all its horrific destruction shall be poured upon the world. Then, the bottomless pit shall be opened. Out of which shall come towering billows of smoke, and such a horde of demonic creatures that shall darken the skies by their sheer number. Those creatures will have one purpose only: to torment humans. They will be everywhere, in every building and space. For fully five months, no one will be able to come out of his hiding place, for fear of being exposed to countless, painful scorpion-like strikes by those hellish creatures.

The Bible says that the suffering will be so severe that men will "seek death"—though, in that time, they will not be able to die! What, then, can be supposed, regarding the mental state of those who will be subjected to such living torments? Many, no doubt, shall be driven to insanity.

But worse even than that awaits them. For, what shall happen to those demonic creatures, after the five months are ended? The Bible does not *explicitly* say what shall happen to them. To be sure, those creatures will not meekly return into the bottomless pit. But there is perhaps a clue in the story of "Legion" (Mark chapter 5).

"Legion": Mass demonic possession

Thousands of devils did for a while inhabit that one man. When they were cast out of him, they begged Christ not to send them into the abyss (the bottomless pit). They were given permission, by Christ, momentarily to enter into a herd of pigs. Yet, even the pigs immediately afterwards rushed into the sea and were drowned. What then happened to those devils who were so briefly in the pigs? The answer is found in yet another request those devils made, of Christ:

> *"My name is Legion: for we are many.* And he [that is, the devils] **besought him much** [literally, *begged* Christ] that he would not send them away **out of the country**.

Now there was there nigh unto the mountains a great herd of swine feeding. And all the devils besought him, saying, *Send us into the swine, that we may enter into them.* And forthwith Jesus gave them leave [permission]. And the unclean spirits went out, and entered into the swine: and the herd ran violently down a steep place into the sea, ([the swine] were about two thousand;) and were choked in the sea.

"And they that fed the swine fled, and told it in the city, and in the country. And they went out to see what it was that was done. And they come to Jesus, and see him that was possessed with the devil, and had the legion, sitting, and clothed, and in his right mind: and they were afraid. And they that saw it told them how it befell to him that was possessed with the devil, and also concerning the swine. And they began to pray him [Jesus] to **depart** out of their coasts."

MARK 5:1-17

The devils begged Jesus not to make them leave that area of the country. Why? Because, those devils knew they had opportunity (for habitation) within the people of that area. When those people heard how that Jesus delivered that man, they did not receive Christ—but they demanded that Jesus go away and leave them alone!

Where did the devils go after the pigs drowned themselves? Those devils, evidently, then went looking for a new dwelling place—that is to say, **embodiment**...*within those same people who told Jesus he was not welcome among them.* Whereas one man had been possessed and tormented by that 'legion' of devils; those same devils, afterwards, must have 'possessed' many individuals then living in that vicinity.

The foregoing was a somewhat lengthy explanation. Yet it speaks to the question earlier posed, namely: What shall happen to those demonic creatures that come out of the bottomless pit, after the five months are ended? Untold multitudes of humans, around the world, shall have been driven mad by the longstanding torment of those hellish creatures. After the five months are ended, the world will then be ripe, so to speak, for mass demon-possession to occur throughout the world. The demonic spirits that emerged from the pit, with locust-like bodies, will probably abandon those forms and take possession of new bodies—that is, human ones.

Mass demon-possession also appears to be connected with something the Bible describes as the fall of "Babylon" (more about that in a later chapter).

The "Destroyer" ascends from the bottomless pit

The locust-creatures are not the only entities confined in the bottomless pit. What other kinds of beings are imprisoned there? The Bible has a lot to say about a certain one of those:

> "And they had tails like unto scorpions, and there were stings in their tails…. And **they had a king over them, which is the angel of the bottomless pit**, whose name in the Hebrew tongue is <u>Abaddon</u>, but in the Greek tongue hath his name **Apollyon**."
>
> REVELATION 9:10-11

Abaddon and Apollyon both mean "Destroyer". That same being—which is called "the angel of the bottomless pit" (a fallen angel), is the "king" over the horde of locust-scorpion creatures. Later on, that same being will kill the Two Witnesses:

> "And when they [the two witnesses] shall have finished their testimony, <u>the beast that **ascendeth out of the bottomless pit**</u> shall make war against them, and shall overcome them, and kill them."
>
> REVELATION 11:7

That "beast", the Destroyer from out of the bottomless pit, cannot be Satan himself. In Scripture, Satan never appears to be in the bottomless pit, until he is confined there at the end of the day of the LORD. Satan and his angels are presently occupying an unseen realm connected somehow with earth's atmospheric heavens:

> "And **there was war in heaven**: Michael and his angels fought against the dragon; and the dragon fought and his angels, and prevailed not; neither was their place found any more in heaven. And <u>the great dragon was cast out</u>, that old serpent, called <u>the Devil, and **Satan**</u>, which deceiveth the whole world: <u>he **was cast out into the earth, and his angels were cast out with him**</u>."
>
> REVELATION 12:7-9

Lucifer cannot then be Apollyon. But Apollyon—the angel of the bottomless pit; the king over the locust-demons; the Destroyer: may very well be—

The Antichrist

It is widely supposed that Antichrist will be a global peacemaker, a charismatic world leader; that he will rule over a united world-government; that he will come from Europe, or from some Middle Eastern country; and, so on, and so on….

But none of those suppositions agrees with Scripture. In the first place, Antichrist will definitely not be human. Here's the proof:

> "And I saw the beast [Antichrist], and the kings of the earth, and their armies, gathered together to make war against him that sat on the horse [Christ], and against his army. And **the beast** [Antichrist] **was taken, and with him the false prophet** that wrought miracles before him, with which he deceived them that had received the mark of the beast, and them that worshipped his image. **These both were cast alive into a lake of fire burning with brimstone**."
>
> REVELATION 19:19-20

At the Battle of Armageddon, Jesus Christ will cause both "the beast" (Antichrist) and the False Prophet to be taken and, *immediately*, those two shall be cast alive into the Lake of Fire—"which is the second death" (Revelation 21:8). In the whole of Scripture, it appears that Antichrist and the False Prophet are the first ever to be cast into the Lake of Fire. More significant, is the fact that, unlike every other *unsaved* human that has ever lived or shall live, Antichrist and the False Prophet shall be cast into the Lake of Fire—*without first appearing before God at the Final Judgment* (the "white throne" Judgment: Revelation 20:11-15). A thousand years more must then pass, before that anyone else—including Satan himself, will be cast into the Lake of Fire. This seems to prove conclusively that the Antichrist and the False Prophet are not human. Nowhere in Scripture is there the least suggestion that Antichrist is a human being.

Antichrist is a fallen angel — not a human being.

The world will not recover

The world shall in no way or measure return to anything resembling governable—much less well-ordered, civil Society, during the day of the LORD. The Judgment of the world—which shall commence with a solar-system-wrecking event (the solar nova), whereby hundreds of millions will perish; and, which, Judgment, shall next release throngs of wicked spirits including scorpion-like creatures arising in great billows of smoke from out of the bottomless pit; and which, Judgment, shall then cast down Satan and his angels unto the earth, in terrible wrath.... How, does anyone suppose, shall such maimed and crippled, terrorized, starving, tormented human beings then remaining on earth—carry on with their lives? much less reorganize Society?

They shall not recover themselves! They did not want the knowledge of God; much less did they want God's Spirit and Word to rule in their hearts and minds. They wanted hell and the devil and the world, to satiate their selfish desires. So, hell and the devil they shall have—until they are utterly consumed.

The notion, of a limited, survivable, even a sufferable period of 'Tribulation'—nay, even a time of great Revival during a season of global peace!—stands in stark contradiction to a right understanding of the day of God's wrath and the Judgment of the world.

Many 'supernatural' things will occur during the day of the LORD, which will destroy this present world.

Israel's Opportunity

"[A]ND THOU SHALT COME UP AGAINST MY PEOPLE OF ISRAEL,
AS A CLOUD TO COVER THE LAND; IT SHALL BE IN THE LATTER
DAYS, AND I WILL BRING THEE AGAINST MY LAND, THAT THE
HEATHEN MAY KNOW ME, WHEN I SHALL BE SANCTIFIED IN
THEE, O GOG, BEFORE THEIR EYES."

EZEKIEL 38:16

There is purpose in all that God does. God's purpose, in the day of the LORD, entails much more than pouring out his wrath upon the world and the devil. God's Kingdom is eternal and vast. God's plan and his purpose involving humanity and the earth, are part of his will concerning his greater Kingdom. Understanding something of God's manifold purpose, is necessary in order to rightly understand the day of the LORD.

➤ Judgment of the world

The solar nova is not some freakish accident of Nature. It is God's doing. Though it is but one of the means whereby God, in the day of the LORD, is going to put down all rebellion, by "destroy[ing] the sinners" out of the world—and not, as many wrongly and dangerously suppose, 'save the sinners':

"Behold, the day of the LORD cometh, **cruel** both with wrath and fierce anger, to lay the land desolate: and he shall **destroy the sinners** thereof out of it."

ISAIAH 13:9

➤ Overthrow of Satan's kingdom

For God's own reasons—at least some of which he has revealed in his Word, Satan and "his angels" currently occupy a realm somehow intimately connected with the earth, such that the devil is called, in Scripture, *"the prince of the power of the air"*

(Ephesians 2:2). The devil and his otherworldly cohorts will try to resist God, and dominate humanity, during the day of the LORD. But God will crush the powers of darkness and put an end to the devil's dominion. (This topic will be discussed in greater detail in a later chapter.)

➤ Honor and reward the Saints

Not everything that God will do during the day of the LORD, is for the purpose of damnation. The Rapture is the first of many ways that God is going to honor and reward the Saints, from the very beginning of the day of the LORD. The then-glorified Saints are going to be greatly more involved in those events on earth, during the day of the LORD, than is commonly supposed (more about that in a later chapter).

➤ Fulfill his promises concerning Israel

God keeps his promises and his Covenants—including those he made long ago, concerning Israel. This present Church Age will culminate in the Rapture of the Church, at the beginning of the day of the LORD. From which time and forward, God will specially deal with the Jews—among whom He will protect and redeem a "remnant" of Jacob's descendants. God's Covenantal dealings with Israel are centrally important to God's purpose, during the day of the LORD.

The Rapture of the Church; the judgment of the world and of Satan's kingdom; and, Israel's long-awaited turning unto the Lord: all together mark the end of this present Age and the beginning of God's New Order on earth. There is therefore much that will be accomplished in the few years comprising the day of the LORD. The remainder of this chapter provides an overview of God's dealings with Israel, during that time.

Ezekiel's War—beginning of the day of the LORD

The regathering of Israel to become a nation—which did occur in 1948, is one of the most important signs indicating the fast-approaching end of the Church Age, and the soon return of Jesus Christ. Centuries before Christ, the prophet Ezekiel outlined the future history of Israel, wherein (Ezekiel chapters 37-39 inclusive) he foretold Israel's dispersal amongst the nations, following their rejection of Christ. Ezekiel next prophesied that very many years after Israel's dispersion, God would regather Israel from all the places where they had been scattered around the world and bring them back to inhabit their ancient homeland, where God would then bless them and cause them to flourish in the land of Israel.

Ezekiel furthermore prophesied that, in the last days, a coalition of militarily strong nations—which Ezekiel identified, would come to make war against Israel. Israel will then have few if any allies that will stand with her. Massively out-numbered and out-gunned by the attacking coalition of forces, Israel will appear to be faced with certain destruction. But in his Vision, Ezekiel saw that God himself will intervene to save Israel:

> "Thus saith the Lord GOD; Behold, I am against thee, O Gog, the chief prince of Meshech and Tubal: and I will turn thee back, and leave but the sixth part of thee, and will cause thee to come up from the north parts, **and will bring thee upon the mountains of Israel**: and I will smite thy bow out of thy left hand, and will cause thine arrows to fall out of thy right hand. **Thou shalt fall upon the mountains of Israel, thou, and all thy bands, and the people that is with thee**: I will give thee unto the ravenous birds of every sort, and to the beasts of the field to be devoured. Thou shalt fall upon the open field: for I have spoken it, saith the Lord GOD. And **I will send a fire on Magog, and among them that dwell carelessly in the isles**: and they shall know that I am the LORD. **So will I make my holy name known in the midst of my people Israel; and I will not let them pollute my holy name any more**: and the heathen shall know that I am the LORD, the Holy One in Israel.

In the above passage, God said he will not only decimate the invading horde. But he will also "send a fire" upon those nations that remain indifferent to Israel's danger. God said he will thus make his holy name known "in the midst of my people Israel", and the heathen shall recognize that God is in the midst of Israel.

But Ezekiel saw something else. He saw the very weapons that God is going to use to deliver Israel, and punish the nations:

> "And it shall come to pass at the same time when Gog shall come against the land of Israel, saith the Lord GOD, that **my fury** shall come up in my face. For in my jealousy and in the fire of **my wrath** have I spoken, Surely in that day there shall be **a great shaking** in the land of Israel; so that the fishes of the sea, and the fowls of the heaven, and the beasts of the field, and **all creeping things that creep upon the earth**, and *__all the men that are upon the face of the earth__*, **shall shake at my presence, and the mountains shall be thrown down, and the steep places shall fall, and *every wall* shall fall to the ground**. And I will call for a sword against him throughout all my mountains, saith the Lord GOD: every man's sword shall be against his brother. And I will plead against him with **pestilence** and with blood; and <u>I will rain upon him, and upon his bands, and upon the many people that are with him, an</u> **overflowing rain**, and **great hailstones, fire**, and **brimstone**. Thus will I magnify myself, and sanctify myself; and I will be known in the eyes of many nations, and they shall know that I am the LORD."
>
> EZEKIEL 38:18-23

A massive, global earthquake—so intense, in fact, "the mountains shall be thrown down"; and, "every wall shall fall to the ground"; and, "all the men that are upon the face of the earth, shall shake". Besides which, God is also going to send down an "overflowing rain, and great hailstones, fire, and brimstone". **It is the solar nova!**

God is going to use the solar nova to deliver Israel, in Ezekiel's War!

In a previous chapter of this book, it was suggested that the solar nova could perhaps destroy the North and South American continents. But the solar nova is going to be a global event. Ultimately, every continent and nation on earth shall suffer greatly its effects. In the earliest days following the solar nova, however, not all places on earth shall be equally affected. Evidently, those geographic areas surrounding Israel—which at that very moment shall be occupied by armies coming to destroy Israel, shall be severely impacted. Though Israel shall largely be spared. For, God is able to direct with pinpoint accuracy his own weapons of war. He will control the scope and severity of the incoming lethal bombardments from the solar nova.

Ezekiel's War and the Rapture
When I published <u>The Seven Seals in Prophecy and in History</u>, in 2014, I did not then understand the nature of those events that comprise the Cosmic Sign (black sun, blood-moon, etc.). All I knew at that time, was, that the Rapture is going to happen at the same time as those cosmic events occur.

That time now being so near at hand; God has evidently been pleased to reveal, in his Word, the intimate association of the Rapture with Ezekiel's War and the solar nova events.

No one will be able to know, by listening to the news media, that a solar nova is about to occur in a few months or years—much less, in a few days or even weeks. But everyone will be able to know, by closely observing international news reports, when that Ezekiel's War is getting ready to happen. In other words, God has provided the knowledge, in his Word, whereby:

We can know the time of the Rapture—within days, or at least within a few weeks, of its occurrence.

Keep a close eye on Israel. When Iran, Russia, Turkey, and certain north African nations begin to form an alliance and express their mutual belligerence toward Israel—which, as of the time of this writing, has already begun to take shape: Ezekiel's War will very soon follow. Yes, we are actually that close to the Rapture!

Cleanup after Ezekiel's War

Scripture provides a wealth of detailed information, regarding the cleanup after Ezekiel's War:

> "And they that dwell in the cities of Israel shall go forth, and shall **set on fire and burn the weapons**, both the shields and the bucklers, the bows and the arrows, and the handstaves, and the spears, and **they shall burn them with fire seven years**: so that **they shall take no wood out of the field, neither cut down any out of the forests; for they shall burn the weapons with fire**: and they shall spoil those that spoiled them, and rob those that robbed them, saith the Lord GOD.

> "And it shall come to pass in that day, that I will give unto Gog a place there of graves in Israel.... And **seven months** shall the house of Israel be burying of them, that they may **cleanse the land**.... And they shall sever out men of continual employment, passing through the land to bury with the passengers those that remain upon the face of the earth, to cleanse it: after the end

of seven months shall they search. And the passengers that pass through the land, <u>when any seeth a man's bone, then shall he set up a sign by it, till the buriers have buried it</u>...."

EZEKIEL 39:9-16

If the reason for burning the weapons was merely to clean up the mess; then, why will it take seven **years** to do that—when it will only take seven **months** to bury the dead and thus "cleanse the land"?

Besides, why burn the weapons at all? Why not keep and use those *as weapons*? How were it more efficient or practical to disassemble manufactured armaments in order to salvage the wood that is in those, rather than utilize virgin wood from trees?

Better still is the question: Why burn wood at all? Within the past few years, vast deposits of oil and natural gas have been discovered within Israel's borders.   Salvaging wood from weapons, *for fuel*, over the course of seven years (nearly the entire duration of the day of the LORD), doesn't make any sense at all...*unless*...a significant proportion of Israel's trees—as well as the infrastructure needed to drill, store, refine, and distribute gas and oil products—were severely damaged or destroyed by the solar nova.

Contaminated bodies of the dead (alluded to in the above-quoted Scripture), in Ezekiel's War, will not be caused by nuclear fallout or by chemical or biological weapons—as many suppose. The solar nova itself shall contaminate the earth, with highly toxic substances. Toxic fallout will also be produced by massive firestorms.

Such evidences and inferences drawn therefrom, as just discussed, further reinforce the suggestion that the solar nova shall result in primitive conditions resembling the Iron Age.

Israel's spiritual turning point

God's deliverance of Israel, in Ezekiel's War, will result in a great spiritual awakening amongst the Jews:

"Behold, I am against thee, O Gog, the chief prince of Meshech and Tubal: and I will turn thee back, and leave but the sixth part of thee, and will cause thee to come up from the north parts, and will bring thee upon the mountains of Israel: and I will smite thy bow out of thy left hand, and will cause thine arrows to fall out of thy right hand. Thou shalt fall upon the mountains of Israel, thou, and all thy bands, and the people that is with thee: I will give thee unto the ravenous birds of every sort, and to the beasts of the field to be devoured. Thou shalt fall upon the open field: for I have spoken it, saith the Lord GOD. And I will send a fire on Magog, and among them that dwell carelessly in the isles: and they shall know that I am the LORD. <u>So</u> **<u>will I make my holy name known in the midst of my</u>** **<u>people Israel</u>**; and **<u>I will not let them pollute my holy</u>** **<u>name any more</u>**: and the heathen shall know that I am the LORD, the Holy One in Israel.

"Behold, it is come, and it is done, saith the Lord GOD; <u>this is the day whereof I have spoken</u>….

"And it shall come to pass in that day, that I will give unto Gog a place there of graves in Israel, the valley of the passengers on the east of the sea: and it shall stop the noses of the passengers: and there shall they bury Gog and all his multitude: and they shall call it, *The valley of Hamongog*. And seven months shall the house of Israel be burying of them, that they may cleanse the land. Yea, all the people of the land shall bury them; and it shall be to them a renown the day that I shall be glorified, saith the Lord GOD. And they shall sever out men of continual employment, passing through the land to bury with the passengers those that remain upon the face of the earth, to cleanse it: after the end of seven months shall they search. And the passengers that pass through the land, when any seeth a man's bone, then shall he set up a sign by it, till the buriers have buried it in the valley of Hamongog. And also the name of the city shall be *Hamonah*. Thus shall they cleanse the land....

"And I will set my glory among the heathen, and all the heathen shall see my judgment that I have executed, and my hand that I have laid upon them. ***So the house of Israel shall know that I am the LORD their God from that day and forward....***

"Therefore thus saith the Lord GOD; **Now will I bring again the captivity of Jacob, and have mercy upon the whole house of Israel**, and will be jealous for my holy name; after that they have borne their shame, and all their trespasses whereby they have trespassed against me, when they dwelt safely in their land, and none made them afraid. When I have brought them again from the people, and gathered them out of their enemies' lands, <u>and am sanctified in them in the sight of many nations</u>; **then shall they know that I am the LORD their God**, which caused them to be led into captivity among the heathen: but I have gathered them unto their own land, and have left none of them any more there. **<u>Neither will I hide my face any more from them: for I have poured out my spirit upon the house of Israel, saith the Lord GOD</u>**."

EZEKIEL 39

The lengthy passage, above, unambiguously declares that God's deliverance of Israel, in Ezekiel's War, will mark the definitive spiritual turning point for the Jews. One sentence, in particular, could not be clearer on that point:

> "So the house of Israel shall know that I am the LORD their God *from that day and forward*."

From what day? From the day when God destroys Israel's enemies, by the solar nova. The "house of Israel" shall know it was God that did it. And because of the timing and the manner whereby that great deliverance shall come, Israel will also know that God did it—*for them*.

The Jews—and Israel

> "For I would not, brethren, that ye should be ignorant of this mystery, lest ye should be wise in your own conceits; that <u>blindness in part is happened to Israel, until the fulness of the Gentiles be come in</u>. And <u>so all Israel shall be saved</u>...".
>
> ROMANS 11:25-26

Throughout the Church Age, the Jews, broadly speaking, have been blinded to the reality of the gospel of Jesus Christ. But when the "fulness of the Gentiles be come in"—in other words, when the last convert from among the Gentiles shall have come to Christ: then, Israel's partial blindness will be taken away. When will that be? Right after the Rapture of the Church.

In the above-quoted Scripture from the book of Romans, the Apostle Paul declared: "And so all Israel shall be saved." The prophet Ezekiel also wrote, that *following* God's deliverance of Israel, in Ezekiel's War:

> "Neither will I hide my face any more from them: for I have **poured out my spirit** upon the house of Israel, saith the Lord GOD."

Does that mean that every Jew then living on earth, after the solar nova and the Rapture, will be converted to Christ? No, it does not. It simply means that, at that time, God will pour out his Spirit upon the Jews. Thus, God will open wide the door of his mercy and grace unto them. But that does not guarantee that every Jew will therefore *receive* God's mercy and grace, by accepting Jesus Christ. For, has not God likewise

71

poured out his Spirit unto the Gentiles, during the Church Age? Yet, they did not all receive God's Spirit.

There will doubtless be many Jews who will persist in their resistance to the Holy Ghost—notwithstanding, the <u>special visitation of God, which the Jews, alone, amongst the world then under judgment, shall be granted at that time</u>. Alas, the Bible seems to indicate that, during the day of the LORD, many of the Jews will fall under the judgment and wrath of God, together with the unbelieving Gentiles. Though not all of the Jews will fall.

What did the Apostle Paul mean, then, where he wrote, that "all Israel shall be saved"? Paul understood—better, perhaps, than did any other Apostle, the distinction made by the Holy Ghost, between "Israel" and the "Jews":

> "For **they are not all Israel, which are of Israel**: neither, because they are the seed of Abraham, are they all children: but, *In Isaac shall thy seed be called*. That is, **they which are the children of the flesh, these are not the children of God**: but the children of the promise are counted for the seed. For this is the word of promise, *At this time will I come, and Sara shall have a son*."
>
> ROMANS 9:6-9

Clarification: "They are not all Israel [spiritually] which are of Israel [that is, descendants of Jacob]: neither, because they are the seed [natural offspring] of Abraham, are they all children [of God]: but, in Isaac [the promised "seed" which typifies Christ] shall thy seed be called". Children of the flesh [whether Jew or Gentile] are not the children of God: but the children of the promise [the promise is Christ] are counted for the seed".

Where Paul wrote, that "all Israel shall be saved"; his reference was to the "Israel of God"—which, phrase, Paul used in another of his writings (Galatians 6:15-16):

> "For **in Christ Jesus neither circumcision** [Jews] **availeth any thing, nor uncircumcision** [Gentiles], **but**

72

a new creature [one 'born again' in Christ]. **And as many as walk according to this rule, peace be on them, and mercy, and upon the Israel of God**."

The explanation just given is not what has been called, "Replacement Theology". No, the Church has not taken the Jews' place in the plan and purpose of God. God has not cast aside the Jews. Paul (by the Holy Ghost) went to great lengths to explain the wisdom of God in his dealings with both Jews and Gentiles: showing, how that God, in his wisdom, could—and does—show mercy unto all.

Even though God does show mercy unto all, yet God has given to each and every person the power of volition (power to choose). And, so, each one must choose—for himself or herself alone, what he or she will do with Jesus Christ. So must each and every one of the Jews make his own choice, during the day of the LORD.

God is merciful unto all. But God's mercy unto the Gentiles (all non-Jews)—which he has abundantly extended to them throughout the Church Age, will no longer be extended to those who, having heard and yet rejected the gospel of Christ, shall then come under God's judgment.

144,000 natural Jews are 'sealed'

"And I saw another angel ascending from the east, having the seal of the living God: and he cried with a loud voice to the four angels, to whom it was given to hurt the earth and the sea, saying, *Hurt not the earth, neither the sea, nor the trees, till we have sealed the servants of our God in their foreheads*. And I heard the number of them which were sealed: and **there were sealed an hundred and forty and four thousand of all the tribes of the children of Israel**. Of the tribe of Juda were sealed twelve thousand. Of the tribe of Reuben were sealed twelve thousand. Of the tribe of Gad were sealed twelve thousand. Of the tribe of Aser were sealed twelve thousand...".

Revelation 7:2-4

It is clear, from the given context, that those **144,000** are natural Jews. Those same individuals are later depicted in *the Revelation*, as being faithful followers of Jesus Christ. But if that be the case, the question then arises: Why will they not also be taken up together with the Church, at the Rapture? The answer is, because, those, **144,000** Jews, <u>are not yet converted to Christ, at the time of the Rapture of the Church</u>.

Nevertheless, God—knowing beforehand that those individuals will receive the gift of God's Spirit, which, after the Rapture of the Church, will be poured out upon the Jews: God will divinely "seal" them for their protection.

Temple-worship begins

By far the majority of Jews are not religious. Far from it. A significant minority of Jews, however, are intensely religious. To those Jews, the construction of a Temple on what they believe is the "Temple Mount" in Jerusalem, is a matter of supreme importance. In fact, they believe the building of a Temple, and the establishment of Temple-worship, has much to do with the coming of their Messiah.

The solar nova will create conditions that will enable a Temple to be built in Jerusalem. The solar nova will not only destroy Israel's enemies, but it will completely reset geopolitical realities in the Middle East.

The religious Jews in Israel will avail themselves of that opportunity, to be sure. Though the Temple they will build in that time, and for reasons earlier explained, will not be anything like the architectural marvel they have long been working to prepare for its construction.

The secular Jews will not oppose a ritualistic system of Temple-worship in Jerusalem at that time and under those circumstances. Unbelievers are not much troubled by spiritually dead religion with its rituals.

The reality of the living Christ, on the other hand, is very offensive to them that do not believe. And just at that same time, Israel is going to experience what the Bible likens to a spiritual resurrection ("life from the dead" Romans 11:15)....

The Sixth Trumpet judgment

The Fifth Trumpet judgment (opening of the bottomless pit), which unleashes the horde of locust-demon creatures, shall suddenly cease at the end of five months. The Sixth Trumpet judgment (the second "woe") shall then begin. The Trumpet judgements are expressly indicated to be sequential in time:

> [**Fifth Trumpet** judgment:] "And they had tails like unto scorpions, and there were stings in their tails: and their power was to hurt men five months. And they had a king over them, which is the angel of the bottomless pit, whose name in the Hebrew tongue is Abaddon, but in the Greek tongue hath his name Apollyon.

> [**Sixth Trumpet** judgment:] "One woe is *past*; and, behold, **there come two woes more *hereafter***. And the sixth angel sounded, and I heard a voice from the four horns of the golden altar which is before God, saying to the sixth angel which had the trumpet, *Loose the four angels which are bound in the great river Euphrates.*

> "And the *four angels* were loosed, which were prepared for an hour, and a day, and a month, and a year, for **to slay the third part of men**. And the number of the army of the horsemen were two hundred thousand thousand [200 million]: and I heard the number of them. And thus I saw the horses in the vision, and them that sat on them, having breastplates of fire, and of jacinth, and brimstone: and the heads of the horses were as the heads of lions; and out of their

mouths issued fire and smoke and brimstone. By these three was the third part of men killed, by the fire, and by the smoke, and by the brimstone, which issued out of their mouths. For their power is in their mouth, and in their tails: for their tails were like unto serpents, and had heads, and with them they do hurt.

"<u>And the **rest** of the men which were not killed by these plagues **yet repented not** of the works of their hands</u>, that they should not worship devils, and idols of gold, and silver, and brass, and stone, and of wood: which neither can see, nor hear, nor walk: **neither repented they** of their murders, nor of their sorceries, nor of their fornication, nor of their thefts."

REVELATION 9:10-21

It seems very unlikely that the above passage depicts a Third World War involving human soldiers who (then without modern hi-tech weaponry or technology, or fuel) will nevertheless slaughter a third of the world's population—beginning at a time merely six months or so after the solar nova! Such massive undertaking (*involving **200 million** soldiers*) would require unprecedented capabilities and capacities in order to mobilize and supply and direct such forces. That would be an unheard-of feat even under the most favorable circumstances. Almost certainly, that will not be possible to any nation or coalition of nations, following the solar nova.

The Sixth Trumpet judgment describes "horsemen" wearing breastplates of fire and brimstone, and riding "horses" with heads "like lions" and having tails "like unto serpents" with heads. Strange creatures, indeed! Is that the Apostle John's best attempt to describe modern-day battle tanks? Or is it, rather, an accurate description of what he saw in his vision?

Four fallen angels are directly responsible both for the mobilization of that extremely large army, and for the ensuing slaughter.

The result, in any case, will be the destruction of *a third of the world's population* at that time.

The Two Witnesses

But that is not all that happens in connection with the Sixth Trumpet judgment. The **Two Witnesses** shall appear about the same time that the bottomless pit is opened, or shortly after the conclusion of that judgment (five months later). When Apollyon shall appear in the world, so, too, will appear a couple of other, far more important individuals...sent by God.

God will send two great prophets to Israel, to shepherd the Jews, as it were, for a season (about three and a half years), during which time God will pour out his Spirit upon the Jews. Many Jews will then receive Jesus Christ; at least **144,000** of them. The Two Witnesses will feed and protect Christ's 'flock' in Israel, like lambs in a sheepfold. But that is not all that the Two Witnesses will do.

The Two Witnesses are identified, in Scripture, as themselves being part of the second "woe" signified by the Sixth Trumpet:

> "And I will give power unto <u>my **two witnesses**</u>, and they shall prophesy a thousand two hundred and threescore days [three and a half years], clothed in sackcloth. These are the two olive trees, and the two candlesticks standing before the God of the earth. And **<u>if any man will hurt them, fire proceedeth out of their mouth, and devoureth their enemies</u>**: and if any man will hurt them, he must in this manner be killed. **<u>These have power</u>** <u>to shut heaven, that it rain not in the days of their prophecy: and have power over waters to turn</u>

them to blood, and to **smite the earth with all plagues, as often as they will**."

"And when they [the Two Witnesses] **shall have finished their testimony**, the beast that ascendeth out of the bottomless pit shall make war against them, and shall overcome them, and kill them…. And they that dwell upon the earth shall rejoice over them, and make merry, and shall send gifts one to another; because **these two prophets tormented them that dwelt on the earth**. And after three days and an half the Spirit of life from God entered into them, and they stood upon their feet…. And they ascended up to heaven in a cloud; and their enemies beheld them. And the same hour was there a great earthquake, and the tenth part of the city fell, and in the earthquake were slain of men seven thousand….

"**The second woe is past**; and, behold, the third woe cometh quickly."

REVELATION 11:3-14

"The second woe is **past**"…when the Two Witnesses are resurrected and ascend to heaven: that is, after they shall have ceased to "torment them that dwelt on the earth". Throughout the previous three and a half years, those two individuals were empowered by God:

"[T]o shut heaven, that it rain not in the days of their prophecy: and have power over waters to turn them to blood, and to smite the earth with all plagues, as often as they will."

REVELATION 11:6

Contrary to what is believed and taught by many, the Two Witness will not be Evangelists to the world. The Bible *explicitly denies* the occurrence of anything resembling Revival among the Gentiles—much less worldwide Revival—during the day of the LORD. One important instance of that crystal-clear denial, is specifically associated with the exact timeframe of the Two Witnesses' ministry:

> "And **the rest of the men which were not killed** by these plagues [the Sixth Trumpet judgment] **yet repented not** of the works of their hands, that they should not worship devils, and idols of gold, and silver, and brass, and stone, and of wood: which neither can see, nor hear, nor walk: **neither repented they** of their murders, nor of their sorceries, nor of their fornication, nor of their thefts."
>
> REVELATION 9:20-21

In the above-quoted passage, the phrase: "...the **rest of** the men **that were not killed** by these plagues", means, virtually everyone who **survived** the mass destruction associated with the Sixth Trumpet Judgment (the 200 million-strong army from the pit). What about those survivors? They "**repented not** of the works of their hands, that they should not worship devils".

There will be no soul-saving Revival among the Gentiles, during the day of the LORD.

In addition to smiting the earth "with all plagues as often as they will", the Two Witnesses, during the three and a half years of their ministry, will stand, in some measure at least, for the protection of Israel. Antichrist will not be able to take possession of the then recently built Jewish Temple, nor will he be able to wreak havoc in Israel, until after the Two Witnesses shall have finished their work as God has ordained. Throughout the time of their ministry, Antichrist will not be able to overcome them.

But protecting Israel from Antichrist does not necessarily mean that the Jews will altogether escape the judgments and plagues that the Two Witnesses will bring upon the earth. Not all the Jews will embrace the grace of God and the message of Jesus Christ. Nor will all of the Jews receive God's messengers. Presumably, then, the unbelieving Jews will also join in the worldwide celebration, when Apollyon kills the Two Witnesses.

Who are the Two Witnesses? Are they Moses and Elijah, as some believe? Perhaps. The Biblical description strongly suggests they are two real individuals, that is, two *mortals*—who do not appear on earth until sometime after the Rapture. Therefore, it seems unlikely that they are Moses and Elijah. For, how could anyone who was a partaker of the Resurrection, afterwards be killed? Anyways, their identity seems less important than their ministries.

Antichrist—a peacemaker?

From the very beginning of the day of the LORD, God will powerfully shake heaven and earth, as his "fury [will be] poured out like fire" (Nahum 1:6), throughout the duration of that incomparably awful time.

Many suppose that Antichrist will somehow initiate or at least facilitate a season of peace—whether for Israel specifically, or for the world at large. But do the following passages mean that Antichrist will be a peacemaker?

> "And in the latter time of their kingdom, when the transgressors are come to the full, a king of **fierce countenance** [Antichrist], and understanding dark sentences, shall stand up. And his power shall be mighty, <u>but not by his own power</u>: and **he shall destroy wonderfully**, and shall prosper, and practise, and **shall destroy the mighty and the holy people**. And through his policy also **he shall cause craft** [deceit] **to prosper in his hand**; and he shall magnify himself in his heart, and **<u>by peace shall destroy many</u>**…".
>
> DANIEL 8:23-25

The prophet Daniel further wrote:

> "And he [Antichrist] shall **confirm the covenant** with many for one week: and in the midst of the week he shall cause the sacrifice and the oblation to cease, and **for the overspreading of abominations** he shall make it desolate, even until the consummation, and that determined **shall be poured upon the desolate**."
>
> DANIEL 9:27

The above-quoted passages reveal that Antichrist will "confirm" some sort of "**covenant**"—with "many for one week"; and, that "by **peace** [he] shall destroy many". But in the midst of that "week" (which a close study of Scripture indicates is a 'week of years', that is to say, seven years), Antichrist will then abrogate that covenant.

Antichrist (whose name, Apollyon, means "Destroyer") will be singularly determined to subjugate humanity. Though his scheming and hatred will specially focus on destroying the Jews. It has long been a favorite fantasy—of what I have called, "mainstream prophecy teaching", that Antichrist is going to make a peace treaty involving Israel, at the beginning of the so-called 'Tribulation'; by which he will seek to present himself as Israel's ally and protector. It is even supposed by many, that Antichrist will negotiate terms whereby Israel will then be allowed to build a Temple in Jerusalem. It is suggested that Antichrist will thus endear himself to the Jews, and by those means will persuade them to receive him as their Messiah.

But in view of the comprehensive teaching of the Bible—especially, pertaining to world conditions during the day of the LORD, that interpretation just mentioned seems untenable, for the following reasons:

1. Israel's enemies are destroyed by the solar nova

At the very beginning of the day of the LORD, Israel's avowed enemies shall be destroyed by the solar nova and by what other means God will choose (i.e., infighting

81

amongst the invaders). After which, Israel will remain (in context) as a regional power. Whom should Israel then feel they need to be protected from?

2. **Antichrist arises from the pit – AFTER Ezekiel's War**
Antichrist cannot be supposed to make a peace treaty with Israel before Ezekiel's War, because, that creature remains shut up in the bottomless pit until sometime after Ezekiel's War.

3. **The solar nova will create opportunity to build the Temple**
The solar nova—and not Antichrist—will create conditions that shall enable the Jews to build a Temple. Soon after the solar nova event, the Two Witnesses will then appear and, during their ministry, Temple-worship will proceed unhindered in Jerusalem. It should be recognized, however, that by far the majority of Jews are not religious.   The Temple-builders and Temple-worshipers will consist of a small but determined minority of the Jews.

4. **Antichrist is withstood by the Two Witnesses**
The Two Witnesses will appear at very nearly the same time as Apollyon emerges out of the bottomless pit. The Two Witnesses will protect the nation of Israel from Antichrist, for exactly three and a half years. During which time, God will pour out his Spirit unto the Jews and gather a harvest of souls from amongst them—including, at least, the 144,000.

5. **It is "the time of Jacob's Trouble"—not Israel's peace**
That seven-year period bounded, apparently, by the timeframe of Israel's covenant with Antichrist, is referred to, in Scripture, as "the time of Jacob's Trouble" (Jeremiah 30:7), and not Israel's peace. When

the Bible uses the name Jacob, in relation to the nation of Israel, it refers to the natural descendants of Jacob, the Jews.

"The word that came to Jeremiah from the LORD, saying, Thus speaketh the LORD God of Israel, saying, Write thee all the words that I have spoken unto thee in a book. For, lo, the days come, saith the LORD, that I will bring again the captivity of my people Israel and Judah, saith the LORD: and I will cause them to return to the land that I gave to their fathers, and they shall possess it. And these are the words that the LORD spake concerning Israel and concerning Judah. For thus saith the LORD; We have heard a voice of trembling, of fear, and not of peace. Ask ye now, and see whether a man doth travail with child? wherefore do I see every man with his hands on his loins, as a woman in travail, and all faces are turned into paleness? Alas! for that day is great, so that none is like it: it is even **the time of Jacob's trouble**; but he shall be saved out of it. For it shall come to pass in that day, saith the LORD of hosts, that I will break his [Antichrist's] yoke from off thy neck, and will burst thy bonds, and strangers shall no more serve themselves of him: but they shall serve the LORD their God, and David their king, whom I will raise up unto them. Therefore fear thou not, O my servant Jacob, saith the LORD; neither be dismayed, O Israel: for, lo, I will save thee from afar, and thy seed from the land of their captivity; and Jacob shall return, and shall be in rest, and be quiet, and none shall make him afraid. For I am with thee, saith the LORD, to save thee: though I make a full end of all nations whither I have scattered thee, yet will I not make a full end of thee: but I will **correct thee in measure**, and will **not leave thee altogether unpunished**."

JEREMIAH 30:1-11

Within a few weeks or months, at most, after the initial solar nova event, the *unbelieving* Jews will enter into some sort of "covenant" involving Apollyon. Throughout the next three and a half years after making that covenant, Apollyon will not interfere with Israel's internal affairs, including their religious practices—because, of the irresistible power of the Two Witness then protecting Israel.

But will Apollyon represent himself as Israel's protector? The context suggests not. Israel's principal (declared) enemies shall very recently have been destroyed by the solar nova. And the fact, that the Jews will erect a Temple in Jerusalem and institute Temple-worship, suggests that the Jews will not then feel constrained to appease the world, as they now do.

It may be that Apollyon merely agrees (covenants) not to molest Israel. Such could be the extent of that "peace" which the Jews may suppose they have attained. But that same agreement, not long afterwards, will be their undoing. For, meanwhile as the Jews embrace Apollyon's promise(s), those *same* Jews will *at the same time* reject the dire warnings of true messengers (the Two Witnesses) sent to them by God. Scripture laments Israel's "covenant with death" and their "agreement with hell"—in the context of an "overflowing scourge" that will soon afterward "pass through" Israel (Isaiah 28:14-18).

Apollyon kills the Two Witnesses

All during the three and a half years of the Two Witnesses' ministry to Israel, there will be very great suffering throughout the earth. Widespread catastrophes and unprecedented hardships will continue, as the result of the then still ongoing effects of the solar nova. Mankind will at the same time suffer such torments and mass killings by otherworldly creatures from the bottomless pit, as cannot even be imagined. But that is not all. For, in addition to all of that, the Two Witnesses shall have the power to "smite the earth with all plagues, as often as they will"!

The Bible scarcely intimates the nature and extent of Apollyon's activities during that time period; though it is certain that Apollyon will be very much engaged in directing the Satanic horde and their ongoing killing-campaigns. Yet, Scripture reveals much concerning Apollyon's activities *after* the first three and a half years are completed. The focus of God's dealings with those then remaining on earth, will shift to a new and different phase: one that will bring Apollyon much more to the fore, in terms of the final outworking of God's purpose.

At the end of the Two Witnesses' ministry, events will transpire such that will enable Apollyon, at last, to make war against and murder the Two Witnesses. Whose deaths mark a critically important turning point in the day of the LORD.

Israel's Calamity

"WOE TO THE INHABITERS OF THE EARTH AND OF THE SEA! FOR
THE DEVIL IS COME DOWN UNTO YOU, HAVING GREAT WRATH,
BECAUSE HE KNOWETH THAT HE HATH BUT A SHORT TIME."
REVELATION 12:12

The previous chapter concluded with the death of the Two Witnesses. But how did that come about—seeing, that Apollyon was unable to harm those Two Witnesses, until their ministry to Israel was completed? Before addressing that important question, it will first be meaningful to explore another important question, namely: Why will many of the Jews embrace Apollyon as their Messiah?

Antichrist—and the "strong delusion"

The Bible plainly teaches that, after the Rapture of the Church, God himself is going to send "strong delusion"—in order that all those who rejected Christ will instead believe a lie and, so, they will be eternally damned:

> "And then shall that Wicked [Antichrist] be revealed, whom the Lord shall consume with the spirit of his mouth, and shall destroy with the brightness of his coming: even him, whose coming is after the working of Satan with all power and signs and lying wonders, and with all **deceivableness** of unrighteousness <u>in them that perish; because they received not the love of the truth, that they might be saved</u>. And *for this cause* **God shall send them strong delusion**, that they should believe a lie: <u>that **they all might be damned who believed not the truth**</u>, but had pleasure in unrighteousness."
>
> 2 THESSALONIANS 2:8-12

86

Antichrist will be the focal-point of that "strong delusion". During the first three and a half years or so after his liberation from the pit, Apollyon will seek to subvert and dominate humanity by means of great deception—including a strange kind of "peace", that is, one that kills:

> "And in the latter time of their kingdom, <u>when the transgressors are come to the full</u>, a **king** of **fierce countenance**, and understanding dark sentences, shall stand up. And **his power shall be mighty**, but not by his own power: and he shall <u>destroy wonderfully</u> [Apollyon: "Destroyer"], and shall prosper, and practise, and shall <u>destroy the mighty and the holy people</u>. And through his policy also **he shall cause craft** ["deception"] **to prosper in his hand**; and he shall magnify himself in his heart, and **by peace shall destroy many**: he shall also stand up against the Prince of princes; but he shall be broken without hand."
>
> DANIEL 8:23-25

The above quoted passage portrays Antichrist as a monster: not a cartoonish sort with fangs and claws, but a powerful, supernatural being obsessed with murderous intent and employing devious means. His grand deception will almost certainly require that he should appear in human-like form (as angels are shown, in Scripture, as being capable of doing).

Antichrist will give the appearance of making peace—all the while slaughtering millions.

"King" of the locust-demons

In an earlier discussion related to the mystery of the locust-creatures' sudden disappearance at the end of five months, the possibility of mass demon-possession was suggested. Yet, there may be another important factor related to the sudden disappearance of those creatures.

The kingdom of Darkness is not organized and ruled by love and respect, but by fear and intimidation. It is a realm wherein the strong dominate and control the weak. The locust-demons are under Apollyon's command. The Bible makes a point of identifying Apollyon as the "king" over those stinging creatures out of the pit. Are those creatures not then subject to their king?

Apollyon will appear as though he put a stop to that plague of locust-demons.

Perhaps, he will appear to 'slay' those myriad creatures, by the power of his word. In that same instance, those demonic creatures would abandon their locust-like bodies in favor of taking up new 'residence'—in multitudes of humans that were driven to the brink of insanity because of that plague.

Because that no human would be capable to perceive the reality of such a situation, the whole world would therefore acknowledge Apollyon as being a great miracle-worker (which, Scripture says, he will be). After all, the locust-demon plague, which <u>Apollyon himself will direct</u>, shall cause such immense suffering that many will seek to die, in order to escape that plague.

The Scripture, above quoted, declares that Antichrist *"by peace shall destroy many"*. What if the "peace" he brings, is, that he stops the locust-demon plague? <u>Under the circumstances, that would be the very kind of **peace** the world would then want most.</u>

Make no mistake, Apollyon is the Destroyer. Those locust-creatures were under his command from the time they were loosed out of the pit. But the Devil and his angels are masters of deception. Still, an even greater deception will be forthcoming.

Antichrist 'saves' the world

After the locust-demon plague is ended, the world will then be confronted by even more devastating judgments—as the Two Witnesses appear. Throughout the next three and a half years after they arrive upon the scene, the Two Witnesses will have authority from God to "smite the earth with all plagues, as often as they will" (Revelation 11:6).

Apollyon will no doubt try to resist those Two Witnesses, though he will not for a while be able to overcome them. But after three and a half years have passed—that is, just about the same time that Satan and his angels shall be cast down to earth, Satan himself will then give (or, in view of Satan's fresh defeat, he will *relinquish?*) his own power and authority to Apollyon:

> "And there was **war in heaven**: Michael and his angels fought against the dragon; and the dragon fought and his angels, and prevailed not; neither was their place found any more in heaven. And **the great dragon was cast out**, that old serpent, **called the Devil, and Satan**, which deceiveth the whole world: he **was cast out into the earth, and his angels were cast out with him**."
>
> REVELATION 12:7-9

> "[A]nd **the dragon** [Satan] **gave him** [Antichrist] **his power**, and his seat, and great authority…. And they **worshipped** the dragon **which gave power unto the beast**: and they **worshipped the beast**, saying, *Who is like unto the beast? who is able to make war with him?"*
>
> REVELATION 13:2-4

After Apollyon receives power and authority from Satan, following Satan's defeat in the heavens, that same Apollyon will then kill the Two Witnesses. **The whole world will wildly celebrate for joy**—even by sending gifts to each other, when the Two Witnesses are killed:

89

"And when they [the Two Witnesses] shall have finished their testimony, the **beast** that **ascendeth out of the bottomless pit** [Apollyon] **shall make war against them, and shall overcome them, and kill them**. And their dead bodies shall lie in the street of the great city, which spiritually is called Sodom and Egypt, where also our Lord was crucified. And they of the people and kindreds and tongues and nations shall see their dead bodies three days and an half, and shall not suffer their dead bodies to be put in graves. And **they that dwell upon the earth shall rejoice over them, and make merry**, and shall **send gifts** one to another; **because these two prophets tormented them** that dwelt on the earth."

REVELATION 11:7-10

Once again, that same Apollyon shall have managed to bring about a certain kind of "peace on earth". Not in the form of a peace treaty or by any political means, but by putting a stop (temporarily) to some of humanity's suffering—that is to say, humanity's suffering . . . of God's judgments.

The whole world—including the *unbelieving* Jews (both secular and religious), will attribute very great honor and give ready acceptance to Apollyon, their great Champion. He delivered the world from the stinging-locust plague. He bravely contended against two ragtag-appearing zealots (the Two Witnesses) who, somehow and for so long, caused so many plagues and inflicted so much torment. Now, Apollyon has managed to kill them.

But to certain devoutly religious Jews in Israel, Apollyon will be so much more. He will be to them the Messiah.

Antichrist is welcomed in Jerusalem

It is well known that many Jews are not at all religious. During the day of the LORD, the unbelieving Jews, together with the rest of the world, will gladly embrace Apollyon—not

90

only because of his *apparent* ability to 'save humanity' from unnatural plagues, but also for his willingness to persecute religious 'zealots', whom the world despises. And for a while now, there has been a great deal of 'zealotry' in the streets of Israel, on the part of the Two Witnesses and their sizeable band of followers.

When Apollyon emerges from the pit, his hatred for humanity will for a season be cloaked by deception. Meanwhile, his contempt for *religion*—particularly, his hatred of the Two Witnesses and their followers (the **144,000** elect Jews), will be condoned and not merely tolerated by the world.

The *unbelieving* Jews will welcome Apollyon, when he comes to possess Jerusalem.

Jesus rebuked the Jews in his day, because they rejected him. In his rebuke to them, Jesus alluded to a future time when the Jews would welcome a false Messiah:

> "I am come in my Father's name, and ye receive me not: if another shall come in his own name, **him ye will receive**."
>
> JOHN 5:43

The crisis of the 144,000

From the beginning of the day of the LORD, and throughout the next three and a half years following, God will pour out his Spirit upon the Jews. The Two Witnesses will appear, whose powerful ministry will lead many Jews (at least **144,000**) to receive Jesus Christ. At the same time, the Two Witnesses shall "smite the earth with all plagues, as often as they will"; and, so, protect the Jews from harm, meanwhile as the Spirit of God is at work among them.

91

At the same time that many Jews will be converted to Christ, however, the *unbelieving* Jews will become increasingly agitated by what they (have always) perceived as being an anti-Jewish cult: that is, a *Christian* one. More important, their own rejection of the Spirit of God will burn like an oven in their souls.

The unbelieving Jews, including the unrepentant Temple-worshipers, will rejoice when the Two Witnesses are killed.

Then, the **144,000** Jewish converts shall promptly face another crisis. Their prophets and pastors (the Two Witnesses)—after their dead bodies shall have lain in the streets for three days or so, shall afterwards be resurrected and taken up into heaven...*without* the **144,000**!

All the while their fellow (unbelieving) Jews are welcoming this powerful miracle-worker, Apollyon, as their Leader. What are the new disciples of Jesus Christ to do?

They won't have long to wait, before they know what they must do. As celebrations in Israel very soon focus on preparations being made...involving the Temple.

The flight of the Jews

The flight of the believing Jews will occur right after that an image to the beast has been placed in the Temple in Jerusalem. Which shall occur very soon after that Apollyon has been welcomed in Jerusalem, as Israel's Deliverer. That will be the sign and the signal for them to flee by way of a valley—one newly formed by what will probably be attributed to a powerful earthquake. Although, in reality, it will be an earthquake caused by the Lord Jesus Christ standing upon the Mount of Olives—significantly, a place *outside* the city of Jerusalem. Jesus himself will come to rescue the 144,000; but not in Jerusalem.

In his prophetic warning to the Jews *in his own time*, Jesus told them to flee when they saw Jerusalem **surrounded with armies**. That prophecy was fulfilled when Jerusalem was destroyed by the Romans in A.D. 70.:

"And when ye shall see **Jerusalem compassed with armies**, then know that the desolation thereof is nigh. Then let them which are in Judaea **flee to the mountains**; and let them which are in the midst of it depart out; and let not them that are in the countries enter thereinto. For these be the days of vengeance, that all things which are written may be fulfilled.

"But woe unto them that are with child, and to them that give suck, in those days! for there shall be great distress in the land, and wrath upon this people. And **they shall fall by the edge of the sword**, and shall **be led away captive into all nations**: and **Jerusalem shall be trodden down of the Gentiles, until the times of the Gentiles be fulfilled**."

LUKE 21:20-24

In the prophecy just quoted, the Jews were instructed to watch for armies that would encircle Jerusalem: which, upon seeing, they should flee to the mountains. That same prophecy also warned that many Jews would be killed, and many others "led away captive into all nations". Clearly, that prophecy, recorded in the gospel of Luke, was fulfilled in every detail, in the first century A.D., in the conquest of Jerusalem, by the Romans.

Two other New Testament writers (Matthew and Mark) each recorded a prophecy seemingly identical to the one in Luke's gospel (above discussed). But close examination of that prophecy, in Matthew and Mark, reveals it is actually a very different warning spoken by Jesus, concerning those <u>Jews who will be living on earth during the day of the LORD</u>:

"When ye therefore shall <u>see the **abomination of desolation**, spoken of by Daniel the prophet, stand in the holy place</u>, (whoso readeth, let him understand:) Then let them which be in Judaea flee into the mountains: let him which is on the housetop not come down to take any thing out of his house: neither let him

93

which is in the field return back to take his clothes.... [F]or <u>then shall be great tribulation, such as was not since the beginning of the world to this time, no, nor ever shall be</u>."

Matthew 24:15-21

Jesus warned the Jews that there would be two different times when they would need to flee Jerusalem. In the first instance, the Jews were instructed to watch for armies encircling the city. But in the second case, the Jews are told to watch for a certain object that will be placed within the Temple in Jerusalem. As it happened, the Temple in Jerusalem was destroyed in A.D. 70. Obviously, the latter prophecy pertains to a still-future time, that is, when an image to Antichrist will be erected in a Temple in Jerusalem.

Those Jews who will be alive at the time when Apollyon kills the Two Witnesses and then comes to take possession of Jerusalem, are **not** instructed to watch out for armies surrounding Jerusalem. Rather, they are instructed to watch for "the abomination of desolation...stand in the holy place".

When that image of Antichrist—which the False Prophet will cause to be created—has been set up within the "holy place" (the innermost compartment, in the Temple); **only then** will those, Jews <u>who believe the prophetic Word of God</u>, know it is time for them to get out of Jerusalem.

Apollyon must already have taken possession of Jerusalem, before that his image will then be set up in the Temple. His conquest of Jerusalem will not be accomplished by force of arms. He will be welcomed by the unbelieving Jews, including many religious (albeit unbelieving) Jews, as their great Deliverer, their Messiah.

Zechariah prophesied a special deliverance

The prophet Zechariah also foretold certain events that are connected with the day of the LORD. Zechariah's prophecy begins by describing the initial assault against Jerusalem at the time of Ezekiel's War. Although God will surely deliver Israel, in that war; yet, Israel shall not be "altogether unpunished" (as Jeremiah wrote):

> "[T]hough I make a full end of all nations whither I have scattered thee, yet will I not make a full end of thee: but I will correct thee in measure, and will <u>not leave thee altogether unpunished</u>."
>
> JEREMIAH 30:11

Zechariah's prophecy thus begins:

> "Behold, **the day of the LORD <u>cometh</u>**, and thy **spoil** shall be **divided** in the midst of thee. For **I will gather all nations against Jerusalem to battle**; and the city shall be taken, and the houses rifled, and the women ravished; and **half of the city shall go forth into captivity**, and the residue of the people shall not be cut off from the city."
>
> ZECHARIAH 14:1-2

The day of the LORD "**cometh**", began Zechariah: the day of the LORD is near at hand—but it is not *yet* come. Zechariah continued: "I will gather all nations against Jerusalem to battle". This cannot then be 'the battle of Armageddon'—which occurs at the <u>conclusion</u> of the day of the LORD. Moreover, none of the Jews will "go into captivity" after the battle of Armageddon.

The prophet Zechariah then announced that the LORD himself will fight to protect and deliver Israel:

> "Then shall the LORD go forth, and fight against those nations, as when he fought in the day of battle."
>
> ZECHARIAH 14:3

All the events of the solar nova are implicated in that one sentence, above. "Those nations" against which the Lord will then fight, are the same nations that he himself did "gather them against Jerusalem to battle"—immediately *before* the day of the LORD begins. When the Lord himself shall fight against those nations, he will not lose the battle. Israel will thus

95

be delivered—at the beginning of the day of the LORD. The event Zechariah next describes shall occur *during* the day of the LORD.

> "And his [Jesus's] feet shall stand in that day upon the mount of Olives, which is before Jerusalem on the east, and the mount of Olives shall cleave in the midst thereof toward the east and toward the west, and there shall be a very great valley; and half of the mountain shall remove toward the north, and half of it toward the south.

> "And **ye shall flee to the valley of the mountains**; for the valley of the mountains shall reach unto Azal: yea, *ye shall flee*, like as ye fled from before the earthquake in the days of Uzziah king of Judah...".
>
> ZECHARIAH 14:4-5

Tremendously significant revelation is contained in the above passages. In order to bring it out more clearly, two statements within those passages are juxtaposed—in order as they appear in the text, as follows:

> "And **his feet shall stand in that day upon the mount of Olives**,...and the mount of Olives shall cleave in the midst...

> "And **ye shall flee** to the valley of the mountains...".

It is very commonly supposed that that passage of Scripture, above, is a reference to the <u>Second Coming of Christ</u>—at the <u>conclusion</u> of the day of the LORD, immediately prior to the battle of Armageddon. But that interpretation cannot be correct. For, the very next verse following, states that the Jews will then "flee to the valley of the mountains".

In the prophetic Scriptures, a lot of emphasis is placed upon, let us call it, the 'flight of the Jews', which will occur at some time *during*, but definitely not after, the day of the LORD. Yet, Zechariah reveals that Christ will stand upon the Mount of

Olives—at a time *before* the Jews take their flight into the wilderness, thru that very same valley created by Christ's cleaving the Mount of Olives.

One of the following must then be true, either:

a) the flight of the Jews from Jerusalem will not occur until the <u>conclusion</u> of the day of the LORD; or,

b) Christ will actually "stand upon the Mount of Olives" at some time much earlier, during the day of the LORD.

Which is it? For, both cannot be true. And the correct interpretation is *necessary* to a right understanding of Bible prophecy. The solution has already been plainly given, where Jesus himself said:

> "When ye therefore shall see **the abomination of desolation,** spoken of by Daniel the prophet, **stand in the holy place**, (whoso readeth, let him understand:) **then** let them which be in Judaea **flee into the mountains...**".

> MATTHEW 24:15-16

The woman in *Revelation* chapter 12

The twelfth chapter of *the Revelation* consists of a detailed and mysterious prophecy related to the time period and events presently under discussion. The main character in that prophecy is a certain "woman". Explanatory comments are interspersed throughout the following text, as noted by an asterisk (*):

> "And there appeared a great wonder **in heaven**; a woman clothed with the sun, and the moon under her feet, and upon her head a crown of twelve stars: and she **being** with child cried, travailing in birth, and pained to be delivered.

> "And there appeared another wonder **in heaven**; and behold a great red dragon, having seven heads and **ten horns**, and seven crowns upon his heads. And his tail drew the third part of the stars of heaven, and did cast

them to the earth: and the dragon stood before the woman which was ready to be delivered, for to devour her child as soon as it was born.

"And she brought forth a man child, who was to rule all nations with a rod of iron: and her child was **caught up** unto God, and to his throne."

REVELATION 13:1-5

* The woman is first seen in the heavens—and not on earth. The woman is triumphant, and crowned. "Clothed with the sun", suggests she is radiant with and filled with light. Whereas, "the moon under her feet", suggests that the shadow, or the mere reflection, of light, is now under her feet. Which intimates she is then in an exalted position and possessed of an exalted consciousness. The entire image is a picture of the triumphant Bride of Christ, in heaven. In that state, and in that place, the dragon cannot touch her.

* The dragon, also then in the heavens (and not yet cast down to the earth), is pictured as having ten horns—but <u>as yet with no crowns on the horns</u>. The horns represent the "ten kings" who will arise to exercise power "with the beast" only when that Antichrist is enthroned. Thus, the prophecy begins at a time prior to Antichrist's dominion.

* There is an allusion to that same woman, in Galatians 4:26: *"Jerusalem which is above is free, which is the **mother** of us all"*. The woman is the "Israel of God": consisting of the redeemed of God: many of whom are even now in glory, though some are still (and until that time) upon the earth.

* The "man-child", as soon as it was born, was "<u>caught up</u>" (Gr. ἁρπάζω - harpazo: "seize, take up by force") unto God and to his throne. Not incidentally, that is the same Greek word that appears in 1 Thessalonians 4:17:

"Then we which are alive and remain shall be **caught up** [harpazo] together with them in the clouds, to meet the Lord in the air: and so shall we ever be with the Lord."

The birth of the "man-child" is a reference to the Rapture of the Church. Continuing, in chapter twelve:

"And the woman **fled into the wilderness**, where she hath a place prepared of God, that they should <u>feed her there</u> a thousand two hundred and threescore days [three and a half years]."

* Very soon after the Rapture (i.e., three and a half years after), the woman then appears on earth, having "fled into the wilderness". This represents the 'flight of the Jews', when the abomination of desolation is set up in Jerusalem. Which also indicates the same time period when that the devil is defeated and cast out of heaven, into the earth—at which time Apollyon was then empowered to kill the Two Witnesses:

"And there was **war in heaven**: Michael and his angels fought against the dragon; and the dragon fought and his angels, and prevailed not; neither was their place found any more in heaven. And **the great dragon was cast out**, that old serpent, called the Devil, and Satan, which deceiveth the whole world: he **was cast out into the earth, and his angels were cast out with him**....

"Woe to the inhabiters of the earth and of the sea! for the devil is come down unto you, having great wrath, because he knoweth that he hath but a short time. And **when the dragon saw that he was cast unto the earth, he persecuted the woman which brought forth the man child**.

"And to the woman were given two wings of a great eagle, that she might **fly into the wilderness**, into her place, where she is **nourished** for a time, and times, and half a time, from the face of the serpent.

"And the serpent cast out of his mouth water as a flood after the woman, that he might cause her to be carried away of the flood. And the earth helped the woman, and the earth opened her mouth, and swallowed up the flood which the dragon cast out of his mouth.

"And the dragon was wroth with the woman, and **went to make war with the remnant of her seed**, which keep the commandments of God, and have the testimony of Jesus Christ."

REVELATION 12:7-17

* The devil, having been cast down to earth, then seeks to destroy the 'woman' that "fled into the wilderness"; but he is hindered from doing so. She is the same woman who earlier gave birth to the man-child—only, when seen in her earth-bound aspect, she is not then pictured as being "clothed with the sun", etc., but she is running and hiding from the devil! Same woman. Different aspect. Different realm. Different members of her Body, and in different circumstances. The Body of Christ is at the same time both in heaven and on earth. There are many saints now in heaven, and many still on earth. At the time which this prophecy points to, the focus is upon that portion of God's redeemed who at that time have "fled into the wilderness", after having seen the sign of the "abomination of desolation", in Jerusalem.

* Most, if perhaps not all, of the 144,000 Jews, then recently converted to Jesus Christ, will "flee into the wilderness", where they will be "fed" and "nourished" there for three and a half years, after that Antichrist comes to power in Jerusalem.

* Unable to molest the Jewish Christians then in hiding and protected by God; the devil (through the agency of Apollyon,) will turn back and "make war against the remnant of [the woman's] seed, which keep the commandments of God and have the testimony of Jesus Christ". It appears that there may still be some number, a remnant, as it were, of Jewish converts

that remain in Jerusalem and its environs. Other passages of Scripture seem to agree with that possibility:

> "And it was given unto him [Antichrist] **to make war with the saints, and to overcome them**: and power was given him over all kindreds, and tongues, and nations."
>
> REVELATION 13:7

> "And in the latter time of their kingdom, when the transgressors are come to the full, a king of fierce countenance, and understanding dark sentences, shall stand up [Antichrist]. And his power shall be mighty, but not by his own power: and he shall destroy wonderfully, and shall prosper, and practise, **and shall destroy the mighty and the holy people**."
>
> DANIEL 8:23-24

The warning, to flee Jerusalem when the abomination of desolation is set up in the Temple, does not appear in the Old Testament. That warning was given by Jesus Christ, and is recorded only in the New Testament—which the unbelieving Jews reject. Similarly, Jesus's warning to the Jews in his own time, to flee Jerusalem when that city would be surrounded by armies (in A.D. 70), was widely known and believed by the Christian Jews, at that time; and they escaped. Whereas, the vast majority of unbelieving Jews perished.

God will judge Israel

> "[A]nd in the midst of the week he [Antichrist] shall cause the sacrifice and the oblation [in Temple-worship] to cease, and **for the overspreading of abominations he shall make it desolate**, even until the consummation, and that determined shall be poured upon the desolate."
>
> DANIEL 9:27

101

"And the king [Antichrist] shall do according to his will; and he shall exalt himself, and magnify himself above every god, and shall speak marvellous things against the God of gods, and shall prosper **till the indignation be accomplished**: for that that is determined shall be done."

DANIEL 11:36

"For, lo, the days come, saith the LORD, that I will bring again the captivity of my people Israel and Judah, saith the LORD: and I will cause them to return to the land that I gave to their fathers, and they shall possess it.

"And these are the words that the LORD spake concerning Israel and concerning Judah. For thus saith the LORD; We have heard **a voice of trembling, of fear, and not of peace**. Ask ye now, and see whether a man doth travail with child? wherefore do I see every man with his hands on his loins, as a woman in travail, and all faces are turned into paleness? **Alas! for that day is great, so that none is like it: it is even the time of Jacob's trouble**; but he shall be saved out of it.

"For it shall come to pass in that day, saith the LORD of hosts, that I will break his [Antichrist's] yoke from off thy neck, and will burst thy bonds, and strangers shall no more serve themselves of him: but they shall serve the LORD their God, and David their king, whom I will raise up unto them....

"For I am with thee, saith the LORD, to save thee: though I make a full end of all nations whither I have scattered thee, **yet will I not make a full end of thee: but I will correct thee in measure, and will not leave thee altogether unpunished**."

JEREMIAH 30:3-11

God's purpose concerning Israel and the Jews, during the day of the LORD, principally has to do with their salvation. Of course, that is God's purpose involving the entire human race. However, the day of the LORD is that time when God will give to the Jews a special season of grace. His Spirit will be poured out to them, throughout the course of three and a half years. It will be a time of great opportunity—and of great testing, for them.

Many Jews will receive Jesus Christ, in that time. Many others will not. Those who refuse Christ will suffer the eternal consequences of that, along with all the other sinners on earth, during the day of the LORD. The judgment that is then coming, during the day of the LORD, will be unlike anything that ever came before.

Antichrist takes control

When Apollyon is enthroned in Jerusalem, his carefully laid trap—baited with so great a deception—will be ready to ensnare the souls of multitudes. At that time, Antichrist will assemble a powerful cabal consisting of ten "kings", to dominate the world. But don't bother trying to identify those kings—which are not humans, but they are part of the Satanic empire. To be sure, they are not contemporary political figures; nor are they related to any ancient political divisions: for, they receive no kingdom—until, the beast comes to power:

> "And the ten horns which thou sawest are **ten kings, which have received no kingdom as yet**; but receive power as kings one hour with the beast. These have one mind, and shall give their power and strength unto the beast."
>
> REVELATION 17:12-13

Which militates against the popular theory that Antichrist's power-base shall be a "revived Roman Empire". His power-base is otherworldly. Apollyon will make men his footstool, not his viceregents.

The False Prophet

There is not going to arise any one-world religion—any more than there shall be a one-world government (of men). Antichrist hates everything having to do with 'religion'. The devil hates everything that humans 'worship' besides the devil. It is the "whore", "Babylon the great", who appears for a little while 'riding' the beast, that revels in idolatry and licentiousness. But the beast's underlords—those ten 'kings' that give their power and strength to the beast:

> "...shall **hate the whore**, and shall make her desolate and naked, and shall eat her flesh, and burn her with fire."
>
> REVELATION 17:16

Many are confused by the fact that the False Prophet shall compel the world to "**worship**" the beast and his image; and, by the fact that Antichrist himself will sit in the Jewish Temple, demanding that he must be worshipped:

> "And I beheld another beast coming up out of the earth [the False Prophet]; and he had two horns like a lamb, and he spake as a dragon. And he exerciseth all the power of the first beast before him [Antichrist], and **causeth** the earth and them which dwell therein to **worship** the first beast, whose deadly wound was healed. And he doeth great wonders, so that he maketh fire come down from heaven on the earth in the sight of men, and deceiveth them that dwell on the earth by the means of those miracles which he had power to do in the sight of the beast; saying to them that dwell on the earth, that they should make an image to the beast, which had the wound by a sword, and did live. And he had power to give life unto the image of the beast, that the image of the beast should both speak, and cause that as many as would not **worship** the image of the beast should be killed."
>
> REVELATION 13:11-15

Beast - worship

The whole issue turns upon the meaning of the word, "**worship**". Worship has nothing at all to do with praise. Those two very different words are oftentimes conflated: as in the expression, *praise and worship*. But they are not at all the same thing. In its simplest definition and clearest meaning:

WORSHIP is: The outward expression of one's inward allegiance.

Worship has nothing, *necessarily*, to do with love or adoration. It has strictly and solely to do with one's *allegiance*, one's *willful submission*. And that is precisely what the devil wants. He could care less whether anyone loves him or not. Of course, love can motivate one to give his or her allegiance to another. But so, too, can fear, or lust. It is the element of *submission* and *obedience* that is the real heart of the matter. *Submit. Obey. Yield.* That—alone—is "worship".

Whether then by temptation and deception—or by tyranny and fear; by all means whatsoever, the devil seeks to *compel* all to submit, obey, and yield to him. That is, to worship him.

Whereas, God, by truth and love, seeks not to coerce, but to *convince* all, by truth and goodness, to submit to him—in order that we may live in righteousness and peace and joy, as God intended. And, thus, give honor and joy to the heart of God our Creator.

There is a real and vital difference between God's purpose and methods, and those by which the devil operates. It is the difference between Life and death.

In either case, however, 'worship' is still *the outward expression of one's inward allegiance.* One's inward allegiance will invariably be expressed in one's choices and conduct. To whom one gives his allegiance and yields himself to serve above all

105

others: *that* one becomes the *object* of worship—and the Master of the worshiper.

The hidden face of idolatry

It can thus be understood how it is that the unsaved world is wholly given to idolatry. For there is one above all others—even above God himself, who is the object of men's worship. That 'idol' is SELF. Unregenerate (unsaved) men's inward allegiance is to their own self-interest. The 'god' they serve above all others, is their own self. The Bible, in many ways, reveals that men are themselves the "gods", the "idols", they worship:

> "For the LORD is great, and greatly to be praised: he is to be feared <u>above all gods</u>. For <u>all the **gods** of the nations **are idols**</u>..."
>
> PSALM 96:4-5

> "I have said, ***<u>Ye are gods</u>***...."
>
> PSALM 82:6

> "Be ye [Christian] not unequally yoked together with **unbelievers**: for what fellowship hath righteousness with unrighteousness? and what communion hath light with darkness? And what concord hath Christ with Belial? or what part hath he that believeth with an infidel? And what agreement hath **the temple of God** [Christians] **with idols** [unbelievers]? for <u>ye are the temple of the living God</u>; as God hath said, I will dwell in them, and walk in them; and I will be their God, and they shall be my people. Wherefore come out from among them, and be ye separate, saith the Lord...".
>
> 2 CORINTHIANS 6:14-17

> "Ye lust, and have not: ye kill, and desire to have, and cannot obtain: ye fight and war, yet ye have not, because ye ask not. Ye ask, and receive not, because ye ask amiss, that ye may consume it upon your lusts. Ye

adulterers and **adulteresses**, know ye not that the friendship of the world is enmity with God? whosoever therefore will be a friend of the world is the enemy of God."

JAMES 4:2-4

Self-seeking. Self-serving. The Holy Ghost, in James, calls them "adulterers" and "adulteresses": harlots: idolaters.

But when Apollyon sits on the throne, that devil will not permit any to resist his authority. The False Prophet will demand that all men must "worship" the beast (Antichrist) and his image. But that does not mean the False Prophet will be a kind of religious cheerleader. He himself is subservient to the stronger entity. The False Prophet (another fallen angel)—long ago bound himself under darkness to the power of the Devil. His coming shall be for one purpose, namely: to *compel* all mankind to give their **allegiance** to the beast [Apollyon]—and, by extension, to the kingdom of Darkness. In other words, the False Prophet will compel all to worship the devil.

Powerful coercion

The False Prophet will have a most powerful weapon wherewith to compel men's submission. He will somehow control access to the basic necessities of life. But his ability to control those resources will neither require a globally-integrated monetary system nor even a global digital infrastructure [including Internet]. As earlier explained, there will be no such technological infrastructure, after the solar nova. But there shall be a great number of demonic entities in the earth at that time— all subject to the ten kings and their master Apollyon: which could monitor and enforce the requirement to worship the Antichrist—as well as control access to food and other resources.

Not only by brutal coercion will the False Prophet lead multitudes (whomsoever receives the "mark" of the beast) to their eternal doom. The False Prophet will also be able—in the presence of the beast (Antichrist) whom he serves, to work great miracles:

> "And he [the False Prophet] doeth great wonders, so that he <u>maketh fire come down from heaven</u> on the earth in the sight of men, and <u>deceiveth them that dwell on the earth by the means of those miracles which he had power to do</u> in the sight of the beast; saying to them that dwell on the earth, that they should make an image to the beast, which had the wound by a sword, and did live. And he had power to give life unto the image of the beast, that the image of the beast should both speak, and cause that as many as would not worship the image of the beast should be killed. And he causeth all, both small and great, rich and poor, free and bond, to receive a mark in their right hand, or in their foreheads: and that **no man might buy or sell, save he that had the mark**, or the name of the beast, or the number of his name."
>
> Revelation 13:13-17

Then, when Antichrist is enthroned, and the whole world has been deceived to embrace him as the one that can lead them out of their imperiled and miserable condition—

And, when the False Prophet, by working great miracles, leads the world to give their allegiance to the devil—

And, when all of Satan's host shall have been brought together on earth, for the Temptation of mankind, in this generation:

Then, the "mark of the beast" shall be imposed upon all. And there will be such an overarching watchfulness, by a determined and supernatural cadre of Satan's minions, that practically the entire world's population (then remaining) will submit to receive the "mark", the visible sign of their eternal allegiance to Antichrist.

Temporal death, the cost of their refusal.
Eternal death, the cost of their submission.

Babylon

"AND CUSH BEGAT NIMROD: HE BEGAN TO BE A MIGHTY ONE
IN THE EARTH. HE WAS A MIGHTY HUNTER BEFORE THE LORD:
WHEREFORE IT IS SAID, EVEN AS NIMROD THE MIGHTY HUNTER
BEFORE THE LORD. AND THE BEGINNING OF HIS KINGDOM
WAS BABEL...IN THE LAND OF SHINAR. OUT OF THAT LAND
WENT FORTH ASSHUR, AND BUILDED NINEVEH...".
                                                  GENESIS 10:6-19

God's judgment of something called, in Scripture, "Babylon the great", is among the most important of all the prophecies in the Bible. And, yet, the true identity and nature of Babylon appears to be among the most impenetrable mysteries in the Word of God—notwithstanding, the considerable volume of text devoted to that subject. But when the layers of that mystery are carefully laid open and its deepest truths are revealed, the reason for God's indignation against Babylon becomes crystal clear.

In the beginning

From the beginning in the Garden of Eden, Satan and his angels have actively worked to subvert humanity—by tempting and deceiving man to put his own interests above all else. In the Old World before the Flood, a race of wicked giants (called, in Scripture, *Nephilim*) was spawned by a number of rebellious angels who took unto themselves human women as wives. By the time of Noah (the eighth generation from Adam), mankind had become so corrupted by sin that God destroyed the whole world, save eight souls (Noah and his immediate family), in the Flood.

Although the Flood drowned all the Nephilim as well as all but those eight human beings, yet the Flood did not drown the devil and his angels—which were just as determined after the Flood, as before, to subvert the human race.

In the second generation after the Flood, one of Noah's grandsons fathered a child, named, Nimrod. Of course, Nimrod never did see or experience, personally, the Old World. All he knew of that world was what stories were told him by those few in his immediate family who came across the Flood. And, oh, what stories those must have been!—including fabulous stories describing the great Cataclysm that destroyed that first world. Nimrod was not the only one among the first generations following the Flood, who heard those stories. Perhaps, the handful of eyewitness accounts of the global Cataclysm somehow influenced Nimrod and others to build cities for their mutual safety. Though it appears that Nimrod—whose name means *"rebellion"*, was largely motivated by pride and personal ambition.

Origins of human self-government

Nimrod asserted himself mightily as a leader in the then newly-emerging world, which at that time was still thinly populated. He not only built cities but he further organized those cities as a "kingdom", over which he did rule. The "beginning of [Nimrod's] kingdom" was a city named Babel—the origin and namesake of Babylon: which, name, means *"confusion; mixing"*. In the heart of Babel, the inhabitants attempted to build a great tower "unto heaven". But for what purpose?

> "[L]et us build us **a city** and a tower, **whose top may reach unto heaven**; and let us **make us a name**, <u>lest we be scattered abroad upon the face of the whole earth</u>."
>
> GENESIS 11:4

Though God had expressly commanded Noah and his sons:

> "And God blessed Noah and his sons, and said unto them, Be fruitful, and multiply, and **replenish the earth**.... And you, be ye fruitful, and multiply; bring forth abundantly in the earth, and multiply therein."
>
> GENESIS 9:1,7

111

One of Nimrod's uncles, named, Asshur, who also was a notable leader in that time period, built the city of Nineveh. Babel (Babylon) and Nineveh were two of the earliest and most prominent cities in the New World after the Flood. The Bible provides few clues—besides those already mentioned, regarding how the devil tempted and deceived many of Noah's early descendants. Yet secular history reveals that those same epicenters of human civilization were also the birthplaces of a host of religious ideas and practices involving the worship—of devils.

Babylon and Nineveh very long ago fell into ruins. Nevertheless, throughout the Bible, Nineveh and, especially, Babylon are used to represent the ancient roots of false religions that have ever since lured the world to engage in idolatrous practices.

Simply put: Babylon was the beginning of organized self-government among men; which *necessarily* was founded upon idolatrous beliefs and practices. The relationship between idolatry and human self-government, shall shortly become clear.

Idolatry—and the worship of devils

In the New Testament, the Apostle Paul summed up all false religious practices as being the worship of devils:

> "What say I then? that the idol is any thing, or that which is offered in sacrifice to idols is any thing? But I say, that **the things which the Gentiles sacrifice, they sacrifice to devils, and not to God**: and I would not that ye should have fellowship with devils. Ye cannot drink the cup of the Lord, and the cup of devils: ye cannot be partakers of the Lord's table, and of the table of devils."
>
> 1 CORINTHIANS 10:19-21

Ultimately, every form of false religious belief and practice is idolatry. Just as the devil first tempted Eve to believe that by means of special knowledge she could become God-like; the devil continues to tempt and deceive multitudes to believe that, by means of certain esoteric knowledge and/or rituals, they can themselves become gods and obtain immortality.

In the vast majority of cases, however, false religious leaders do not expressly lead their followers to worship devils as such. Rather, those "blind leaders of the blind" use their respective religious organizations and teachings, for their own gain:

> "But there were false prophets also among the people, even as there shall be false teachers among you, who privily shall bring in damnable heresies, even denying the Lord that bought them, and bring upon themselves swift destruction. And many shall follow their pernicious ways; by reason of whom the way of truth shall be evil spoken of. And <u>through covetousness shall they with feigned words make merchandise of you</u>...."
>
> 2 PETER 2:1-3

By means of false religions, the devil has all along achieved at least the following three goals that evidently are very important to him, namely:

1) he has been able to seduce and subvert multitudes of humans, from worshipping the true God;
2) he has thus deceived those same individuals to embrace beliefs that further alienate them from the true God;
3) through false religions, the devil receives unto himself— if not expressly then at least tacitly—the worship, of such persons.

With the foregoing discussion in mind, let us see what the Bible has to say about "Babylon"—and God's judgment against that, during the day of the LORD:

> "And there came one of the seven angels which had the seven vials, and talked with me, saying unto me, Come hither; I will shew unto thee the judgment of the great whore that sitteth upon many waters: with whom the kings of the earth have committed fornication, and the

113

inhabitants of the earth have been made drunk with the wine of her fornication. So he carried me away in the spirit into the wilderness: and I saw a woman sit upon a scarlet coloured beast, full of names of blasphemy, having seven heads and ten horns. And the woman was arrayed in purple and scarlet colour, and decked with gold and precious stones and pearls, having a golden cup in her hand full of abominations and filthiness of her fornication: and upon her forehead was a name written, MYSTERY, BABYLON THE GREAT, THE MOTHER OF HARLOTS AND ABOMINATIONS OF THE EARTH.

"And I saw the woman drunken with the blood of the saints, and with the blood of the martyrs of Jesus: and when I saw her, I wondered with great admiration. And the angel said unto me, Wherefore didst thou marvel? I will tell thee the mystery of the woman, and of the beast that carrieth her, which hath the seven heads and ten horns.

"The beast that thou sawest was, and is not; and shall ascend out of the bottomless pit, and go into perdition: and they that dwell on the earth shall wonder, whose names were not written in the book of life from the foundation of the world, when they behold the beast that was, and is not, and yet is. And here is the mind which hath wisdom. The seven heads are seven mountains, on which the woman sitteth. And there are seven kings: five are fallen, and one is, and the other is not yet come; and when he cometh, he must continue a short space. And the beast that was, and is not, even he is the eighth, and is of the seven, and goeth into perdition.

"And the ten horns which thou sawest are ten kings, which have received no kingdom as yet; but receive power as kings one hour with the beast. These have one mind, and shall give their power and strength unto the beast. These shall make war with the Lamb, and the Lamb shall overcome them: for he is Lord of lords, and King of kings: and they that are with him are called, and chosen, and faithful.

"And he saith unto me, The waters which thou sawest, where the whore sitteth, are peoples, and multitudes, and nations, and tongues. And the ten horns which thou sawest upon the beast, these shall hate the whore, and shall make her desolate and naked, and shall eat her flesh, and burn her with fire. For God hath put in their hearts to fulfil his will, and to agree, and give their kingdom unto the beast, until the words of God shall be fulfilled. And the woman which thou sawest is that great city, which reigneth over the kings of the earth."

REVELATION CHAPTER 17

The above-quoted passage of Scripture describes "Babylon" as a whorish woman who "sitteth upon many waters". That woman is also shown riding atop a certain beast. The "many waters", the beast, and the woman, are three different entities. Let's examine each of them in turn, as follows.

"Many waters"

The interpretation of the phrase, "many waters", seems straightforward and simple enough to understand:

"The waters which thou sawest, where the whore sitteth, are peoples, and multitudes, and nations, and tongues."

Whatever that whorish woman, "Babylon", may be, it is something that "sits"—as a "queen" (Revelation 18:7), in some sense, above people throughout the whole world. Furthermore, both the "kings" as well as the "inhabitants" of the earth have all committed "fornication" with that whorish woman. In some vital aspect, then, <u>Babylon is a kind of universal principle common to all mankind—a principle that has corrupted humanity from ancient times</u> (as indicated by the appellation, Babylon), <u>until now</u>.

The "beast"—the system, and its leader

The "beast" upon which the woman rides, is somewhat more difficult, but not too difficult, to understand its meaning. The beast represents both a system as well as a certain leader of that system.   The description of that system is explicitly identified with Satan (which name means "adversary"):

> "And there appeared another wonder in heaven; and behold a **great red dragon**, <u>having seven heads and **ten horns** [without crowns]</u>, and seven crowns upon his heads…. And there was war in heaven: Michael and his angels fought against the dragon; and the dragon fought and his angels, and prevailed not; neither was their place found any more in heaven. And <u>the great</u> **<u>dragon</u>** <u>was cast out, that old serpent,</u> **called the Devil, and Satan**, which deceiveth the whole world…."
>
> REVELATION 12:3,7-9

The beast is clearly Satanic. It has seven "heads" and "ten horns". The seven heads are further defined as representing both seven "mountains" and seven "kings" associated with those mountains:

> "The seven heads are <u>seven mountains</u>, on which the woman sitteth.   And there are **seven kings**: five are fallen, and one is, and the other is not yet come; and when he cometh, he must continue a short space. And the beast that was, and is not, even **he is the eighth, and is *of* the seven**, and goeth into perdition."

History reveals that, in the interval between the Flood and the time when the Apostle John wrote the Revelation, there had been six world empires, viz.: Egyptian; Assyrian; Babylonian; Medo-Persian; Grecian; and Roman. The seventh world empire (Islamic) did not arise until shortly after the dissolution of the Roman empire, several centuries after John.

But the mystery deepens. The Bible says that, during the day of the LORD, there is going to arise—not an eighth "head"—but an eighth "king". One of the "kings" of the seven world empires, is going to be given, as it were, new life. He will then be the "eighth" king:

> "And there are seven kings: five are fallen, and one is, and the other is not yet come; and when he cometh, he must continue a short space. And the beast that was, and is not, even **he is the eighth, and is <u>of</u> the seven**, and goeth into perdition."

There remains a part of the riddle to be solved which has to do with the nature and identity of those seven "kings". In fact, every world empire, above identified, each had numerous kings, during the long history of each of those respective empires. There were not then only *seven* kings, who ruled over the course of something like four-thousand years. There were actually very many more. Why does the Scripture mention only seven?

Earthly kingdoms and evil angels

In the book of Daniel, it is revealed that earthly kingdoms are somehow overruled by (fallen) angels. When the angel Gabriel was dispatched to deliver to Daniel a message from heaven, Gabriel was withstood for fully three weeks, by "the prince of Persia" (Daniel 10:13)—a fallen angel. Gabriel further informed Daniel:

> "[N]ow will I return to fight with the **prince of Persia**: and when I am gone forth, lo, the **prince of Grecia** shall come."

> DANIEL 10:20

Each of the seven world empires was evidently under the powerful influence of one of seven spiritual monarchs—that is to say, fallen angels. That is why the Bible identifies only seven "kings" with those seven "mountains", or empires. And one of those kings is going to return to rule, again—during the day of the LORD.

Do not miss the supremely important point: <u>The reign of Antichrist is not going to be one of human design, nor shall it be of human origin, nor shall it be of human form</u>. It shall be wholly Satanic not only in its nature, but in every way Satanic in fact. The "eighth" "king"—a fallen angel now imprisoned in the bottomless pit, is going to arise from out of that pit to rule over the final, Satanic empire that will exist for a very brief time during the day of the LORD.

The eighth "king" will not then be invisible—as was the case during his former reign. But he will be fully visible to human beings. Angels—and even the Lord himself—have sometimes appeared unto men, in human form; instances of which occurred both at the Resurrection and at the Ascension of Christ, and elsewhere:

> "And they entered in, and found not the body of the Lord Jesus. And it came to pass, as they were much perplexed thereabout, behold, two **men** stood by them in shining garments: and as they were afraid, and bowed down their faces to the earth, they said unto them, Why seek ye the living among the dead?"
>
> LUKE 24:3-5

> "And while they looked stedfastly toward heaven as [Jesus] went up, behold, two **men** stood by them in white apparel; which also said, Ye men of Galilee, why stand ye gazing up into heaven?"
>
> ACTS 1:10-11

And when the Lord appeared unto Abraham:

> "And the LORD **appeared** unto [Abraham] in the plains of Mamre: and he sat in the tent door in the heat of the day; and he lift up his eyes and looked, and, lo, three **men** stood by him: and when he saw them, he ran to meet them from the tent door, and bowed himself toward the ground, and said, My Lord, if now I have found favour in thy sight, pass not away, I pray thee, from thy servant...."
>
> GENESIS 18:1-3

Quite possibly, angels naturally have a human-like appearance. At any rate, the beast that will ascend out of the bottomless pit shall evidently not be the only one, from the world of devils and demons, to operate openly, visibly, on earth, during the day of the LORD. But all the wicked creatures of hell, both from below and from above, shall be given, *by God*, the power to dominate the whole world, openly, manifestly, for a very little while during the day of the LORD. That is why Jesus said:

> "For then shall be great tribulation [trouble], such as was not since the beginning of the world to this time, no, nor ever shall be."
>
> MATTHEW 24:21

The beast system in its final form shall begin to be consolidated, following the opening of the bottomless pit; and be fully expressed when Antichrist will be enthroned at Jerusalem. Antichrist's chief minister, the False Prophet—another fallen angel, will then seduce the world to worship Antichrist, by the astounding miracles which he, the False Prophet, will perform. The False Prophet will also organize and administer a worldwide program whereby every human shall be compelled to give his or her allegiance (worship) to the Antichrist. All who refuse will be killed by beheading.

Antichrist's dominion shall also include "ten kings" (the "ten horns" on the seven heads of the beast), who will "receive power as kings one hour with the beast":

"And the ten horns which thou sawest are **ten kings, which have received no kingdom as yet; but receive power as kings one hour with the beast**. These have one mind, and shall give their power and strength unto the beast. These shall make war with the Lamb, and the Lamb shall overcome them…".

REVELATION 17:12-14

Those ten kings are not human. Antichrist will not look unto man for help, nor humble himself before humans. When the Antichrist rises to power, together with the False Prophet and the Ten Kings: then, the beast—which for millennia past did carry (enabled, supported) that whorish woman (Babylon), will then no longer tolerate that woman:

"And the ten horns which thou sawest upon the beast, **these shall hate the whore**, and shall make her desolate and naked, and shall eat her flesh, and burn her with fire. For God hath put in their hearts to fulfil his will, and to agree, and give their kingdom unto the beast, until the words of God shall be fulfilled."

REVELATION 17:16-17

At last, it appears that **the beast actually hates the whore, Babylon**. For millennia past, the beast was willing to be used by the woman (Babylon) for her own self-glorification. But during the day of the LORD, when the kingdom of darkness shall openly dominate the world, the devil and his legions will no longer tolerate Babylon. And here is a very important key to understand what is Babylon:

Babylon is hated both by God…and by the devil.

When Antichrist comes fully to power, then the judgment of Babylon will be suddenly escalated. It is important, though, to distinguish between the wrath that the devil will unleash against Babylon, and the wrath that God will pour out upon the same. The Scripture just quoted, above, reveals that the beast's 'ten horns' will "hate the whore, and shall **make her desolate and naked,** and shall **eat her flesh, and burn her with fire**".

But the Bible clearly shows that, <u>at a time soon after that,</u> God is going to allow his saints to carry out the ultimate judgment and punishment of Babylon. But how can that be? If the Antichrist and his wicked angel-rulers carry out such seemingly total destruction of Babylon: then, what would remain of Babylon, for the saints to execute judgment upon? We will revisit that question, a little later.

In sum, the beast-empire is the manifestation of Satan's dominion in, and over, the affairs of (fallen) humanity. The beast-system is evidenced in the world-system, which is driven by lust, greed, and self-will; a sensual, materialistic kingdom ruled by fear and intimidation. A kingdom overruled by evil angels.

The woman, Babylon

We have thus far described a number of significant clues related to the identity of Babylon—which the Bible calls, a "mystery" (Revelation 17:5). Yet there are further clues given in the Scriptures:

> "And the woman was arrayed in purple and scarlet colour, and decked with gold and precious stones and pearls, having a golden cup in her hand <u>full of abominations and filthiness of her fornication</u>: and upon her forehead was a name written, MYSTERY, BABYLON THE GREAT, THE MOTHER OF HARLOTS AND ABOMINATIONS OF THE EARTH. And <u>I saw the woman drunken with the blood of the saints, and with the blood of the martyrs of Jesus....</u>"
>
> REVELATION 17:4-6

121

Ever since the beginning, Babylon has displayed a murderous hatred for all who truly love God. Yet, *somehow*, God's own people have themselves been a part of Babylon:

> "And I heard another voice from heaven, saying, **Come out of her** [Babylon], **my people**, that ye be not partakers of her sins, and that ye receive not of her plagues. For her sins have reached unto heaven, and God hath remembered her iniquities."
>
> REVELATION 18:4-5

In the passage of Scripture just quoted, the Word of God is addressed to "my people", that is, to God's people. The following passage is a continuation of that address to the people of God:

> "Reward her [Babylon] even as she rewarded you ["my people"], and double unto her double according to her works: in the cup which she hath filled [with bloodshed] fill to her double. How much she hath glorified herself, and lived deliciously, so much torment and sorrow give her: for she saith in her heart, I sit a queen, and am no widow, and shall see no sorrow. Therefore shall her plagues come in one day, death, and mourning, and famine; and she shall be utterly burned with fire: for strong is the Lord God who judgeth her."
>
> REVELATION 18:6-8

Babylon, the great whore, is called, a "city"

Many believe that the "Babylon" of prophecy is a literal city on earth. The question, then, related to Babylon's identification as a "city" must be resolved. The Bible identifies Babylon as a great city, in the following:

> "Alas, alas, that **great city Babylon**, that **mighty city**! for in one hour is thy judgment come.... And a mighty angel took up a stone like a great millstone, and cast it into the sea, saying, Thus with violence shall that **great**

city Babylon be thrown down, and shall be found no more at all."

In the natural sense, a city is an aggregate of buildings including houses and other types of structures. Those buildings serve but one purpose only, which is, to accommodate the people who live and work in and near unto such a place. But an aggregate of buildings, apart from the souls that inhabit those, does not constitute a city. There a numerous such places on earth—where no one lives, though very many buildings still remain in those places. A city, therefore, is really an aggregate not of buildings, but of persons living in society one with another. Those persons, of course, require buildings of all kinds to facilitate the activities of life. The Bible also alludes to the fact, that a city is an aggregate of persons living in common society:

> "There is a river, the streams whereof shall make glad the **city of God**, the holy place of the tabernacles of the most High. God is in the midst of **her**; she shall not be moved: God shall **help her**, and that right early."
>
> PSALM 46:4-5

> "The LORD loveth the gates of **Zion** more than all the dwellings of Jacob. Glorious things are spoken of thee, O **city of God**. Selah.... And of Zion it shall be said, This and that man was born in her: and the highest himself shall establish her. The LORD shall count, when he writeth up the people, that this man **was born there.**
>
> PSALM 87:2-6

Is Zion, then, an actual city on earth, and the only place whence Christians are born? Or does that "river" of God, and its "streams", only flow in some celestial city made all of gold? Or is it rather the case that Scripture sometimes uses metaphors, including "city", to convey deeper meaning?

123

A city furthermore consists not merely as an aggregate of persons. Cities, as such, tend to outlast generations. Continuity of culture (beliefs and practices) is central to the existence and identity of a city.

In the Biblical sense, then, the term "city" designates a group or society of persons, which have in common certain beliefs and practices, that is to say, a culture.

Babylon is such a society and culture, such a "city".

This mystery, "Babylon", has existed for several thousands of years. It has outlasted many generations of souls. Yet the *culture*—the defining Ideas and beliefs characteristic of Babylon, has persisted unto this day.

There are many who suppose that "mystery" Babylon is a literal, physical city that must first be (re)built and then become the preeminent center of global commerce, *before* the day of the LORD. But that view necessitates that the day of the LORD cannot then occur for yet a very long time to come. That view further implies that, among all the godless cities of the world, Babylon will uniquely incur the wrath of God, during the day of the LORD. Whereas, the wrath of God is going to be poured out upon the whole world, and not uniquely upon any one city, Babylon.

Finally, when Babylon shall have been completely destroyed, the Bible pronounces a most peculiar judgment against her, as follows:

> "Babylon the great is fallen, is fallen, and is become the **habitation** of devils, and the **hold** of every foul spirit, and a **cage** of every unclean and hateful bird."
>
> REVELATION 18:2

Even in Babylon's destruction, that "city" shall continue to be "inhabited" by devils. More than that, Babylon shall forever

124

be made a "hold", a "cage"—that is to say, a prison, for "every unclean and hateful bird"...that is, **a prison for devils**. Finally, Babylon will not be "a" prison but it will be "THE" prison-house; and not of "**some**" devils but of "EVERY" foul spirit and of "EVERY" unclean and hateful bird.

Babylon cannot then be a literal, physical city.

Summary of evidences

Following, is a summary of all the clues above discussed, related to the identity of Babylon the Great:

 - ➢ Babylon is as ancient as history—and still exists today.
 - ➢ All people in the world have committed spiritual 'fornication' with Babylon.
 - ➢ God's people are commanded to "come out of her".
 - ➢ Babylon hates all them that love God, and has literally killed countless numbers of the same.
 - ➢ Babylon places high value on materialism, sensuality, and worldly pleasure.
 - ➢ Babylon is wealthy and politically powerful.
 - ➢ Babylon is profane and blasphemous.
 - ➢ Babylon is hated by God, and by the devil.
 - ➢ Babylon is going to be destroyed by the beast-system.
 - ➢ Babylon is going to be destroyed by the saints of God.
 - ➢ Babylon shall forever be the prison-house of devils.

The **culture** of Babylon is idolatrous, licentious, blasphemous, murderous.

The "**city**" of Babylon is something that will be capable to serve as a veritable <u>prison for all the devils</u>—*after* that Babylon has been destroyed!

Why do God and the devil both hate Babylon?

Looking over the foregoing list of Babylon's defining characteristics, it is very easy to see why God hates Babylon. But it is profoundly perplexing to try to understand why the devil hates Babylon. Yet, the answer becomes perfectly obvious—when the question is cast in a different light, viz.:

What is it that both God and the devil want, involving human beings?

Answer: **They both want the "worship" of man.**

In an earlier chapter, the real nature and meaning of "worship" was discussed in considerable detail. In that same discussion, it was explained that, by reason of his own fallen nature, man is an idolater. However, the **object** of (fallen) man's worship is <u>neither God nor the devil</u>; but, every man's highest **allegiance** (his **worship**) is **to his own self**.

God hates the idolatry of man. But so, too, does the devil. The devil desperately wants man's worship. But understand the reality of what is meant by "worship": The **object** of worship (the one who is worshipped), <u>is the Master of the worshipper</u>. The devil does not merely seek man's downfall and destruction. No. The devil wants to be man's Master. The devil wants the use of man's capacities—especially, man's body.

But as long as man's allegiance and devotion, his worship, is given to his own SELF, the devil has not full control over that individual.

Idolatry (of SELF) is THE defining characteristic of Babylon.

Throughout many centuries past, Babylon has served the devil well. So long as the devil remained in the shadows, Babylon was able to enrich and empower herself by using the world-system (which is Satanic) to her (Babylon's) own advantage. By means of which, the devil was able to advance his dark enterprise among men.

But when the devil comes out of the shadows into the open, during the day of the LORD, he will no longer suffer Babylon to 'ride the beast', as it were. The beast will then seek to utterly destroy 'Babylon'—the overarching principal of **self-worship** and **self-government** that defines fallen humanity. The devil will demand humanity's complete, uncompromising submission to himself, and not to anyone or anything else.

Babylon's two-fold destruction

> "Babylon the great **is fallen**, **is fallen**, and is become the habitation of devils, and the hold of every foul spirit, and a cage of every unclean and hateful bird."
>
> REVELATION 18:2

After that Antichrist rises to power, Apollyon and his evil cabal will seize control of the world's resources. The False Prophet will implement a system of monitoring and control, known in Scripture as "the mark of the beast". It will involve both a *test* of each individual's allegiance to the devil, as well as it shall also incorporate a "seal", in the case of those who pass the devils' test, so to speak.

The test is simple. It goes something like this: "**Give your allegiance to Apollyon. Else, we will kill you.**" Everyone who yields himself, will then receive a certain "seal" upon his body. And they will be granted access to obtain the basic necessities of life. Those who refuse, the Bible reveals, will be beheaded.

But what else will happen to those who pledge their allegiance to the devil? The devil will take absolute control of them in that very same moment, of course: instant—and total—demon possession. The Bible declares that such persons, having thus surrendered themselves to the mastery of Satan, are thenceforth irredeemable.

Thus, the first aspect of Babylon's fall and judgment shall have been accomplished, namely: <u>self-worship shall utterly cease...and with that:</u>

Human self-government shall be forever abolished.

Thus, the beast will destroy the *spiritual aspect*, that is the *culture*, of Babylon. By taking possession of the individual's mind (spirit), the devil will then be Master not only of the 'house', the body. Moreover, the then-indwelling demonic spirit will forever be united with the spirit of that individual. God made man's spirit and body to serve as the Temple of God. <u>Man's spirit somehow has the capacity to be indwelt by, and joined together with, God's Spirit.</u> But those who refuse the Spirit of God...are still capable of being indwelt by, and joined together with, another spirit (or spirits). Demonic ones.

The second aspect of God's two-fold judgment of Babylon involves the destruction of the *physical* aspect, that is, the visible aspect of that "city" called Babylon. That judgment then awaits only the time when all who will receive the mark of the beast shall have done so. At which time God will command the total destruction of Babylon: all those demon-filled "buildings": the bodies and souls of men.

The government of God's Kingdom

God created man to be God's own living Temple. Man is not a robot or a pre-programmed machine. Man is a living soul made in the very image and likeness of God. Man has a mind, a heart, a will, a personality. All by God's will and purpose. God made man with the capacity to perceive God, relate to

God, commune with God: on God's level, in very real and meaningful ways and measure.

But God did never intend for man to live as though he were his own king and Master. Jesus said: "Thine (LORD) is the kingdom, and the power, and the glory forever." The kingdom is the LORD's. God's kingdom is one of peace and love. Which requires that it must then be a kingdom of law and good Order. Man is obliged to live under God's authority and rule; it's as simple as that.

The Fall of man affected man's nature in such a way that he is bent toward prioritizing his own interests above all else—even at the cost of defying the living Spirit and Word of God.

Long ages ago, man's lust to be free from all authority except that of his own will, began to express itself more and more without restraint, until, man began to extend his own will, in effect, over the will and lives of all others.

Human *self-government* was never God's will.

God never sanctioned human self-government. The Institution of human *Civil* Government, which is ordained by God, is required, in the Word of God to serve in honor of, and submission to, the revealed will and ordinances of God:

> "The God of Israel said, the Rock of Israel spake to me, He that ruleth over men **must** be just, ruling in the fear of God."
>
> 2 SAMUEL 23:3

The Rapture of the Church; the closing of the Church Age; the coming of the day of the LORD: signifies the end of the practice of human *self*-government. Forever. Thenceforth, Christ shall sit as King over all the earth (Zechariah 14:9). That is not to say there will never again be rebellion in the hearts of men. But mankind will never again be allowed to rule himself

129

and others, apart from God's authority. All rebellion shall be put down swiftly and without mercy (with a "rod of iron"), from the day of the LORD and onward.

Babylon is fallen. Is fallen. And it shall never rise, again:

> "Thus shall Babylon sink, and ***shall not rise*** from the evil that I will bring upon her...".
>
> JEREMIAH 51:64

The Furnace of Judgment

"AND I SAW ANOTHER SIGN IN HEAVEN, GREAT AND MARVELLOUS, SEVEN ANGELS HAVING THE SEVEN LAST PLAGUES; FOR IN THEM IS FILLED UP THE WRATH OF GOD."
REVELATION 15:1

Nearly four years have now passed (in the context of this book,) since the initial solar nova event occurred. Within the first few weeks, *more than a **billion** souls perished* as the result of a cascade of catastrophic events triggered by the solar nova. Not only have those catastrophic events continued, but their effects have grown worse and worse.

Shortly after the solar nova...and the continent-sized firestorms...and the world-shaking earthquakes...and the destruction of all modern technology-based systems on earth... Shortly after all that, two-hundred million creatures—*of some kind*—destroyed yet another **2 billion** people, or more, during a killing orgy that somehow(?) ended just before that Apollyon was led, with great celebration, into Jerusalem. (Did Apollyon have anything to do with *seeming* to stop that plague, too?)

As if all of that weren't enough to bring the world absolutely to its knees: *Every* fallen angel and demonic spirit—from the abyss below, and from the heavens above—are now gathered together on earth. Two of the most powerful of those evil beings, are openly ruling over the crippled, dying mass of humanity; at least, over what is left of that.

Now (at this point, in this analysis), Apollyon (the Destroyer), and his miracle-working Prime Minister the False Prophet, are enforcing a worldwide operation—whereby Hell's legions are systematically subjugating and 'branding' (with the "mark of the beast"), and at the same time fully "possessing", the then-remaining population of earth.

The entire population of earth (with *very few* exceptions) **is becoming demon-possessed.**

That process (involving testing and possession) will take some time to complete. Though it will take far less time than it might have, only a few years prior. For, at this point in time, now slightly more than halfway thru the day of the LORD, the world's population has been reduced by about half. And those that remain are mostly to be found on the Eurasian and African continents; the entire Western hemisphere having been destroyed by the earliest effects of the solar nova.

This is what Jesus described as "*the hour of temptation* which shall come upon all the world, to try them that dwell upon the earth." Which he also promised his faithful followers he would "keep them from" (Revelation 3:10).

The "Seven Thunders"

Let us briefly go aside to examine an important prophecy that, apparently, was ready for its fulfillment just before, or perhaps during, the time when the Two Witnesses were still on earth (as part of the Sixth Trumpet judgments). Following, is the relevant text:

> "And I saw another mighty angel come down from heaven, clothed with a cloud: and a rainbow was upon his head, and his face was as it were the sun, and his feet as pillars of fire: and he had in his hand a little book open: and he set his right foot upon the sea, and his left foot on the earth, and cried with a loud voice, as when a lion roareth: and when he had cried, **seven thunders** uttered their voices.

> "And when the seven thunders had uttered their voices, I was about to write: and I heard a voice from

heaven saying unto me, **Seal up those things which the seven thunders uttered, and write them not**.

"And the angel which I saw stand upon the sea and upon the earth lifted up his hand to heaven, and sware by him that liveth for ever and ever, who created heaven, and the things that therein are, and the earth, and the things that therein are, and the sea, and the things which are therein, **that there should be time no longer**: but in the days of the voice of the seventh angel, when he shall begin to sound, the mystery of God should be finished...."

REVELATION 10:1-7

There is no point in guessing what kinds of judgments might have been signified by those "seven thunders". That they were judgments, of some kind, seems all but certain. But that John was commanded to *"seal up those things...and write them not"*, is not an insignificant detail. Whereas a casual reading of the text could lead to the conclusion that nothing more can be learned about those seven thunders. Yet, the context in which that prohibition appears, provides an important clue as to the reason why the Spirit of God included that exchange, between John and the heavenly dignitary, in *the Revelation*.

John heard the telling of those events, when the "seven thunders uttered their voices". Those events were really about to happen, evidently. But, then, John was commanded to *"seal them up; write them not"*. In other words, they comprised a set of judgments that were about to manifest—but God intervened to prevent them from ever occurring. Jesus had prophesied that God would "shorten those days", for the sake of the "elect" (the Jewish believers in that time):

"And except that the Lord had shortened those days, no flesh should be saved: but <u>for the elect's sake</u>, whom he hath chosen, **he hath shortened the days**.

MARK 13:20

134

It cannot be a mere coincidence that, after John was commanded not to write the judgments uttered by the seven thunders, the angel immediately then proclaimed that there should be "**time no longer**"; furthermore, that when the next angel would begin to sound his trumpet (the Seventh Trumpet), the *"mystery of God should be finished"*. God would then quickly wrap things up, is the sense.

That curious-sounding phrase, "time no longer", does not mean that time as such would cease to be; as some suppose. Of course, time will not cease to be. Is not the Millennial Kingdom, still then to come, described in Scripture as lasting for "*a thousand years*"? Rather, "time no longer" seems to mean that the seven thunders judgments—*then being set aside*, would therefore cause no further delay.

All things considered, it appears that those verses of Scripture reveal *the manner of God's intervention*, to "shorten those days": <u>by **annulling** a number of judgments</u> that, in lieu of God's intervention, might have served, somehow, to prolong "the elect's" days of suffering.

John's experience, in this part of his Vision, that is to say, in the context of the *sequence of events* that were being revealed to him, occurred near the beginning of the Two Witnesses' ministry. But those events (seven thunders judgments), evidently, were never to be *actualized*—excepting, the 'event' of God's intervention by fiat. And that is important enough for us to perceive and understand: God keeps his promises; God cares for his people.

Jerusalem: the crucible of God's fury

Antichrist will establish the seat of his kingdom in the heart of Jerusalem:

> "And the king [Antichrist] shall do according to his will; and he shall exalt himself, and magnify himself above every god, and shall speak marvellous things against the God of gods, and **shall prosper till the indignation**

135

be accomplished: for that that is determined shall be done....

"And he shall plant the tabernacles of his palace between the seas in the glorious holy mountain...."

DANIEL 11:36;45

The *unbelieving* Jews who only recently embraced Apollyon and gave him the 'keys to the city' of Jerusalem, so to speak, will soon be melted in the fiery furnace of the Destroyer's hideous ferocity. God has laid the wood to the pile. And they in their rebellion have kindled the flame:

"Can thine heart endure, or can thine hands be strong, in the days that I shall deal with thee? I the LORD have spoken it, and will do it. And I will **scatter thee** among the heathen, and disperse thee in the countries, and will consume thy filthiness out of thee....

"And the word of the LORD came unto me, saying, Son of man, the house of Israel is to me **become** dross: all they are brass, and tin, and iron, and lead, in the midst of the furnace; they are even the dross of silver.

"Therefore thus saith the Lord GOD; Because ye are all become dross, behold, therefore I will **gather you into** the midst of **Jerusalem**. As they gather silver, and brass, and iron, and lead, and tin, into the midst of the furnace, to blow the fire upon it, to melt it; so will **I gather you in mine anger and in my fury, and I will leave you there, and melt you**. Yea, **I will gather you, and blow upon you in the fire of my wrath, and ye shall be melted** in the midst thereof. As silver is melted in the midst of the furnace, so shall ye be melted in the midst thereof; and ye shall know that I the LORD have **poured out my fury** upon you."

EZEKIEL 22:14-22

136

In the first part of the prophecy, above, God is seen to **scatter** the Jews among the heathen, and **disperse** them in the countries of the Gentile nations. All for the purpose **to** "consume their filthiness" *out of them*.

But in the next part of that same prophecy, God's complaint is that, rather than being cleansed of their filthiness, they *became* "dross", like the "dross of silver". The refining fires through which they did pass (when they were scattered and dispersed), had not the effect to purify them but, instead, revealed their impurity; for, they (the *unbelieving* Jews) are dross, and not silver. In other words:

They came forth, thru the fires (in their dispersal), **not as silver, but as dross.**

Therefore, the LORD will then "**gather** them—where? "into the midst of **Jerusalem**...into the midst of the furnace". God said he will then:

> "...blow the fire upon it, to melt it; so will I gather you in mine **anger** and in my **fury**, and <u>I will **leave you there,** and **melt you**</u>...and ye shall know that **I the LORD have poured out my fury upon you**".

A similar and related prophecy is found in Isaiah:

> "Wherefore it shall come to pass, that <u>when the Lord hath performed</u> [completed] his **WHOLE WORK** upon **mount Zion and on Jerusalem**, I will [then] punish the fruit of the stout heart of the king of Assyria [a metaphor for Antichrist], and the glory of his high looks....

> "And it shall come to pass in that day, that **the remnant of Israel**, and such as are **escaped** of the house of Jacob, shall no more again stay upon him that smote

137

them; but shall stay upon the LORD, the Holy One of Israel, in truth. The **remnant** shall return, even the remnant of Jacob, unto the mighty God.

"For though thy people Israel be as the sand of the sea, yet a remnant of them shall return: the consumption decreed shall overflow with righteousness. For the Lord GOD of hosts shall make a consumption, even determined, in the midst of all the land.

"Therefore thus saith the Lord GOD of hosts, O my people that dwellest in Zion, be not afraid of the Assyrian: he shall smite thee with a rod, and shall lift up his staff against thee, after the manner of Egypt. For yet a very little while, and **the indignation shall cease**, and mine anger **in their destruction**."

ISAIAH 10:12; 20-25

The "consumption decreed"

Throughout the previous three and a half years, in the days of the Two Witnesses' ministry, God poured out his Spirit upon the Jews then in Israel. Though many Jews did receive the mercy of God, during that season of grace, yet, very many more resisted the Spirit of God in their midst.

God had already, in days prior even to that time, dealt with the unbelieving Jews who were scattered among the nations, in ways that God wanted should "consume their filthiness" out of them. But they refused. Punishments and affliction did not heal their souls. Nor, then, did the Word preached unto them with great power and anointing heal them. What were then left that God could do to save them from their own destructions?

"Ah sinful nation, a people laden with iniquity, a seed of evildoers, children that are corrupters: they have forsaken the LORD, they have provoked the Holy One of Israel unto anger, they are gone away backward. **Why should ye be stricken any more? ye will revolt more and more**: the whole head is sick, and the whole heart faint. From the sole of the foot even unto the head there is no soundness in it; but wounds, and bruises, and putrifying sores: they have not been closed, neither bound up, neither mollified with ointment."

ISAIAH 1:4-6

It is not as though those same Jews did not want a Deliverer. But they wanted to be delivered from their troubles, and not from their own sins. Such is the case with all who love the world; who will not come to the light, because they hate the light, because their deeds are evil.

"I gave her *space to repent* of her fornication; and she repented not."

REVELATION 2:21

"It is a fearful thing to fall into the hands of the living God."

HEBREWS 10:31

Apollyon—so they believed, would be their Deliverer. Ah, the lying devil. The unbelieving Jews then remaining will then discover, to their horror, that Apollyon is no Deliverer: but he is the "rod" of God's anger and the "staff" of his indignation, to take them away, in the fury of God's wrath and, so, will God cleanse his own land.

Men and devils, prepared for destruction

As the then-remaining human survivors come, as if *en-masse,* to trade their eternal souls to the devil and receive his spirit and 'seal', in exchange for a morsel of bread: an indescribable pall of darkness appears to cover them, like a shroud upon the living dead.

In the digital fantasy-world that now exists (in this present generation) by means of computer technologies, including the Internet; fictional creatures, called, "zombies", have been wildly popularized. But in the soon-coming day of the LORD, that fiction ("zombies") will then become an unfathomably dark reality.

But how will God then deal with a world filled with devils and demon-possessed-men? Who will stop Apollyon from extending his dominion over all the earth? How can the comparatively few remaining humans—those who have not yet surrendered themselves utterly to the devil's possession, possibly prevail against such seemingly irresistible forces?

> "And they worshipped the dragon which gave power unto the beast: and they worshipped the beast, saying, Who is like unto the beast? **who is able to make war with him?**"

REVELATION 13:4

Reaping the wheat

Ever since the False Prophet installed the "abomination of desolation" in the Temple at Jerusalem (above three years ago, in this timeline), the believing remnant of the Jews has remained hidden somewhere in the wilderness. For them, doubtless the past seven years or so have been exceedingly difficult. Though it is certain that God will take care of them.

Will there be some martyrs from among their number? God knows. Yet there must be a substantial number of Christ's faithful followers still alive on the earth, when the following occurs, as the end of the day of the LORD approaches:

> "And I heard a voice from heaven saying unto me, Write, Blessed are the dead which die in the Lord from henceforth: Yea, saith the Spirit, that they may rest from their labours; and their works do follow them. And I looked, and behold <u>**a white cloud**</u>, and <u>upon the cloud one sat like unto **the Son of man**</u>, having on his head a golden crown, and **in his hand a sharp sickle**. And another angel came out of the temple, crying with a loud voice to him that sat on the cloud, Thrust in thy sickle, and reap: for the time is come for thee to reap;

140

for the harvest of the earth is ripe. And he that sat on the cloud thrust in his sickle on the earth; and the earth was reaped."

REVELATION 14:13-16

Why would God first "seal" the 144,000 Jews, at the time of the solar nova; and, then, pour out his Spirit upon them, during the course of the Two Witnesses' ministry; after which, Jesus Christ cleaved in two the Mount of Olives, to make a way of escape through the "valley of the mountains", for the 144,000 believing Jews; following which, they fled into the wilderness where they were "fed" and "nourished" <u>for three and a half years</u>: Why would God do all of that, if, at the very end, God intended to give them over to be slaughtered by Apollyon? Antichrist is the instrument of God's wrath. And God said in his Word that the faithful in Jesus Christ have not been appointed unto wrath.

Thank God! when Christ thrusts in his sickle to reap the earth, the **144,000** will be translated and glorified, then to take their place in the heavenly armies of Jesus Christ.

Reaping the tares

The time has at last come to reap the tares, the poisonous weeds, as it were, from out of the earth:

"And another angel came out from the altar, which had power over fire; and cried with a loud cry to him that had the sharp sickle, saying, Thrust in thy sharp sickle, and gather the clusters of the vine of the earth; for her grapes are fully ripe. And the angel thrust in his sickle into the earth, and gathered the vine of the earth, and cast it into **the great winepress of the wrath of God**. And the winepress was trodden without the city, and **blood came** <u>out of the winepress, even **unto the horse bridles**, by the space of a thousand and six hundred furlongs</u>."

REVELATION 14:18-20

141

Whether or not that is supposed to indicate a specific measure of volume of blood, yet it vividly suggests destruction so great that cannot even be imagined. The fact that no geographical location is associated with that bloodbath (except, that it is "without the city"), leaves open the possibility that it encompasses all the earth. Which, in fact, it shall.

This Honor

"DO YE NOT KNOW THAT THE SAINTS SHALL JUDGE THE WORLD? AND IF THE WORLD SHALL BE JUDGED BY YOU, ARE YE UNWORTHY TO JUDGE THE SMALLEST MATTERS? KNOW YE NOT THAT WE SHALL JUDGE ANGELS? HOW MUCH MORE THINGS THAT PERTAIN TO THIS LIFE?"

1 CORINTHIANS 6:2-3

Many object to the idea of the Rapture, for the preposterous reason, as they suppose, that those who believe in the Rapture are just looking for an easy way out. Why should Christians expect they shall escape, via the Rapture, the kinds of suffering that so many others have been subjected to?

True perspective of the Rapture

Those who raise such objection, obviously, do not understand the Rapture. They do not realize that the Rapture is not the great escape; it is the great *equipping*. Jesus Christ is not coming to take his faithful servants out of harm's way. Oh, no. Instead,

---

The Raptured, glorified saints *are the harm* that is coming to this present evil world!

The Rapture will of course spare the true Body of Christ from suffering the wrath of God (not even to mention the wrath of the devil), during the day of the LORD. But we shall not be spared by being removed out of the world. Rather, we shall be protected and preserved from all harm, thenceforth forever, by virtue of our glorification with Christ at his appearing. At the moment of the Rapture, those whom Christ will find them ready and waiting for his appearing, shall be changed: from earthly to heavenly; from merely human, to super-human; from mortal, to immortal—literally, "in the twinkling of an eye" (1 Corinthians 15:52).

The meeting place of the Rapture

Nor is the Rapture a time when the Church is going to fly away to heaven to party while the world burns. As a matter of fact, those who will be Raptured shall not be going very far away from the earth:

> "For the <u>Lord himself shall **descend from heaven**</u> with a shout, with the voice of the archangel, and with the trump of God: and the dead in Christ shall rise first: then we which are alive and remain shall be caught up together with them **in the clouds**, to meet the Lord **in the air**....."
>
> 1 Thessalonians 4:16-17

There is no "air" and there are no "clouds" in outer space. Air and clouds comprise earth's atmospheric 'heavens'. That is where the Raptured saints are going to meet Christ when he comes. Not incidentally, that is also the same realm that Satan and his angels presently occupy. Satan is called, in Scripture, "the prince of the power of the air" (Ephesians 2:2). The devil's spere of cosmic operations, everyone should know, is planet earth. The devil himself is not yet in hell.

Jesus is not coming to secret his sanctified saints away from the bad devil. But Christ is coming to take us with him when, as Captain of the LORD's host, he leads the Archangel Michael and his angels to storm the devil's fortress and kick that squatter out of the heavens—both him and his sinister angels: "Know ye not that **we shall judge angels?**" (1 Corinthians 6:3):

> "And there **was war in heaven**: Michael and his angels fought against the dragon; and the dragon fought and his angels, and prevailed not; neither was their place found any more in heaven. And **the great dragon was**

cast out, that old serpent, called the Devil, and Satan, which deceiveth the whole world: **he was cast out into the earth, and his angels were cast out with him**."

The time of Christ's 'Second Coming'

Is it true, as it has long been supposed and taught by many, that the 'Second Coming' of Christ shall occur at a time very near unto the end of the 'Tribulation': at which time Jesus and his saints will then appear from heaven, riding white horses and coming to destroy the Antichrist and his armies at the battle of Armageddon?

In an earlier chapter in this book, some space was given to resolve a seeming enigma pertaining to the time when that Jesus Christ's own "feet shall stand in that day upon the mount of Olives" (Zechariah 14:4). That important event, it was shown, will occur not at the battle of Armageddon but, rather, at about the same time that the False Prophet will set up an image to the beast, in the Temple at Jerusalem.

The Bible expressly declares that Jesus Christ will "descend from heaven"—*prior to* the Rapture. Whereafter, as Captain of the LORD's host, Christ will then lead his forces including Michael and his angels, in that great war in the region of earth's heavens. After casting the devil down to earth, Jesus will then stand his own nail-pierced feet upon the Mount of Olives, splitting that ancient hill in half, in order to make a way of escape for his Jewish followers who will flee through that newly-created valley into the wilderness (as they are still in the flesh), for a little while.

When, therefore, is the 'Second Coming' of Jesus Christ? The answer is glaringly obvious. The day of the LORD is that time when Jesus Christ himself will judge and punish this present evil world. He is not going to sit on his throne in heaven and watch it unfold via satellite television.

"The LORD is a man of war: the LORD is his name."

EXODUS 15:3

146

Post-Rapture: the then-glorified saints

All the while as Christ leads his heavenly hosts and executes his judgments, during the day of the LORD, the then-glorified saints shall be with him at his side:

> "Then we which are alive and remain shall be caught up together with them in the clouds, to meet the Lord in the air: and **so shall we ever be with the Lord.**"
>
> 1 THESSALONIANS 4:17

But the world will never know that the Rapture did occur. No bodies shall be seen flying up into the sky or rising up from out of the ground. (See my book, **The Seven Seals in Prophecy and in History**, for a more in-depth discussion of the Rapture.) Nothing else but the then-lifeless bodies of those who were Raptured, shall be seen at that time by those on earth. Thus, the Rapture—by its very nature (and by its coincidence with the solar nova), is going to be completely hidden from the world. But not for long.

The saints shall judge angels

The Scripture, quoted at the head of this chapter, reveals that the saints shall judge angels. Of course, it is not the holy angels that shall be judged. But the saints shall judge those, fallen angels, that have rebelled against God. Scripture reveals the precise time when that judgment shall occur:

> "The earth shall reel to and fro like a drunkard, and shall be removed like a cottage; and the transgression thereof shall be heavy upon it; and it shall fall, and not rise again. And it shall come to pass in that day, that the LORD shall punish the host of **the high ones that are on high**, and the kings of the earth upon the earth. And they [the wicked angels] shall be gathered together, as prisoners are gathered **in the pit**, and shall be **shut up in the prison**, and after many days shall they be visited."
>
> ISAIAH 24:20-22

"And I saw an angel come down from heaven, having the key of the bottomless pit and a great chain in his hand. And he laid hold on the dragon, that old serpent, which is the Devil, and Satan, and <u>bound him a thousand years, and cast him into the bottomless pit</u>, and shut him up, and set a seal upon him, that he should deceive the nations no more, till the thousand years should be fulfilled: and after that he must be loosed a little season. <u>And I saw thrones, and they [glorified saints] sat upon them, and judgment was given unto them</u>...".

REVELATION 20:1-4

The day of the LORD is the time when God is going to "punish the host of the high ones that are on high". Satan and his (fallen) angels shall then (in the closing days of the day of the LORD) be "shut up in the prison"—the bottomless pit, for a thousand years then still to come.

Those same angels are the ones who, throughout the ages and by various means, did tempt, torment, and persecute the saints. At a time near the mid-point of the day of the LORD, the devil and his angels shall be cast down out of the heavens into the earth. On his way down, down, down, Satan will give his power and authority to Apollyon. For a very little while.

But why does God allow the devil and his angels any such seeming liberty? Why is Satan not yet bound in the bottomless pit—or cast into the Lake of Fire? The reason is, because, the devil is being used to fulfill God's own purposes, namely: to try the will of each individual, concerning Christ; and, as agents and instruments of the Curse of Sin; and, to demonstrate to all of God's creatures the righteous judgments of God.

In a passage of Scripture in Isaiah, is seen one instance of how that God uses the wicked for God's own purpose:

"O **Assyrian, the rod of mine anger**, and the staff in their hand **is mine indignation. I will send him** against an hypocritical nation, and **against the people of my**

148

wrath will I give him a charge, to take the spoil, and to take the prey, and to tread them down like the mire of the streets. *Howbeit he meaneth not so, neither doth his heart think so*; but it is in his heart to destroy and cut off nations not a few."

ISAIAH 10:5-7

The appearing of Antichrist, during the day of the LORD, in no way signifies that the powers of darkness shall have somehow managed to overtake the world—in defiance of God. Though Antichrist will doubtless think so (as the above passage suggests). Rather, it is God who will **send** Antichrist—"the beast who was, and is not, and yet is" (Revelation 17:8)—**to deceive** and **to tempt** (with death) all who rejected the love of the truth; and, **so, draw them into the net** of God's judgment and wrath.

But not long (about three and a half years) after Antichrist's appearance, God is then going to remove the devil and his angels from amongst the world, and imprison them in the bottomless pit for another thousand years then still to come. Antichrist and the False Prophet shall at that same time be "cast alive" into the Lake of Fire (Revelation 19:20). And God is going to give to his saints—during the day of the LORD, the "honor" of judging *and binding* those fallen angels; as revealed in the following passage of Scripture:

> "Let the **saints** be joyful in glory: let them sing aloud upon their beds. Let the high praises of God be in their mouth, and a twoedged sword in their hand;
>
> ➢ "to execute vengeance upon the heathen, and punishments upon the people;
>
> ➢ "to **bind their kings** with chains, **and their nobles** with fetters of iron;
>
> ➢ "to execute upon them the judgment written:
>
> "**This honour** have all his saints. Praise ye the LORD."

PSALM 149:5-9

The saints have no such prerogative, to execute vengeance and punishments, during this present Church Age. The saints shall indeed "rule the nations with a rod of iron," during the coming Millennial Kingdom Age. Yet, it is *during the day of the LORD*—the time of God's "fierce anger" and vengeance—that **the saints will execute** (carry out) not 'a' judgment, nor yet 'some' judgment, but **"THE judgment written"**. Which, responsibility and privilege, God said in his Word, is an "honor" belonging to "all his saints".

But why shall the saints then **execute** "vengeance" and "punishments" **upon people**; but merely "*bind* their kings and nobles"? Who are those rulers?

Those, overlords of the unbelieving world of human beings, are not themselves human, but they are fallen angels. The saints shall have the "honour" of *judging* and *binding* those, fallen angels, who are now the "principalities" and "powers" and "rulers of the darkness of this world" (Ephesians 6:12). But judging and binding the wicked angels is not all that the saints will do, during the day of the LORD.

The saints shall judge the world

The above-quoted Scripture (Psalm 149) reveals yet another, most important truth, namely:

The then-glorified saints shall be on earth, *during the day of the LORD.*

Not only to judge fallen angels. But the saints shall also, at that same time, execute *God's* "vengeance", by carrying out "punishments" against all those, *humans* then remaining on earth who will have given their allegiance to the Antichrist. God said: "This honour have all his saints. Praise ye the LORD."

But why, should anyone suppose, would God confer that high "honour" upon those, professing Christians, who are not now deeply angered and grieved by the evil that is in this world? Nominal professors of the Christian religion—who, unhappily, constitute the majority of so-called "Christendom", are oftentimes tacitly complicit, if they are not in many cases actively engaged in furthering rebellion against God. The Bible declares what shall be the end of such, hypocrites:

> "Now will I rise, saith the LORD; now will I be exalted; now will I lift up myself.... And the people shall be as the burnings of lime: as thorns cut up shall they be burned in the fire.... The **sinners in Zion** are afraid; **fearfulness** hath **surprised** the hypocrites."
>
> ISAIAH 33:10-14

> "For, behold, the day cometh, that shall burn as an oven; and **all the proud**, yea, and **all that do wickedly**, shall be stubble: and the day that cometh shall burn them up, saith the LORD of hosts.... And ye shall tread down the wicked; for they shall be ashes under the soles of your feet in the day that I shall do this, saith the LORD of hosts."
>
> MALACHI 4:1-3

The phrase, "the day of the LORD", is not a kind of generic euphemism suggesting God's indignation against His enemies in every Age. The day of the LORD is a specific and unique time, during which God is going to judge and punish this present world. God is going to give his then-glorified saints the "honor" of carrying out ("execute") that judgment. Nowhere in Scripture is there a clearer depiction of the saints executing that Judgment, than in the book of Joel:

> "Blow ye the trumpet **in Zion**, and sound an alarm **in my holy mountain**: let all the inhabitants of the land tremble: for **the day of the LORD** cometh, for it is nigh at hand; a day of darkness and of gloominess, a day of

clouds and of thick darkness, as the morning spread upon the mountains: a **great people and a strong**; there hath not been ever the like, neither shall be any more after it, even to the years of many generations. **A fire devoureth before them; and behind them a flame burneth**: the land is as the garden of Eden before them, and behind them a desolate wilderness; yea, and **nothing shall escape them**. The appearance of them is as the appearance of horses; and as horsemen, so shall they run. Like the noise of chariots on the tops of mountains shall they leap, <u>like the noise of a flame of fire that devoureth the stubble</u>, as a strong people set in battle array. Before their face the people shall be much pained: all faces shall gather blackness. They shall run like mighty men; they shall climb the wall like men of war; and they shall march every one on his ways, and they shall not break their ranks: neither shall one thrust another; they shall walk every one in his path: and **when they fall upon the sword, they shall not be wounded**. <u>They shall run to and fro in the city; they shall run upon the wall, they shall climb up upon the houses; they shall enter in at the windows like a thief.</u> **The earth shall quake before them; the heavens shall tremble: the sun and the moon shall be dark, and the stars shall withdraw their shining** [the Cosmic Sign]: and the LORD shall utter his voice before **his army**: for his camp is very great: for he is strong that **executeth** his word: for <u>the day of the LORD</u> is great and very terrible; and who can abide it?"

JOEL 2:1-11

"Let the <u>saints</u> be joyful in glory…. Let the high praises of God be in their mouth, and a twoedged sword in their hand; to execute vengeance upon the heathen, and punishments upon the people; to bind their kings with chains, and their nobles with fetters of iron; to

execute upon **them** <u>**the judgment written**</u>: **This honour have all his saints**. Praise ye the LORD."

PSALM 149:5-9

"Do ye not know that the saints shall judge the world? [....] Know ye not that we shall judge angels?"

1 CORINTHIANS 6:2-3

The prophet Joel described in vivid detail the perfectly thorough manner in which the glorified saints are going to hunt down and destroy all of God's enemies—from city-to-city; from field-to-field; and, quite literally, from house-to-house:

"A fire devoureth before them; and behind them a flame burneth: the land is as the garden of Eden before them, and behind them a desolate wilderness; yea, and nothing shall escape them.... They shall run to and fro in the city; they shall run upon the wall, they shall climb up upon the houses; they shall enter in at the windows like a thief."

JOEL 2:3,7,9

"And ye shall tread down the wicked; for they shall be ashes under the soles of your feet in the day that I shall do this, saith the LORD of hosts."

MALACHI 4:3

Joel's prophecy is not a description of that final conflict that shall take place in open fields in the valley of Megiddo, at the conclusion of the day of the LORD. But Joel shows the saints performing 'search-and-destroy' missions in "cities" and in "houses", running upon "walls" and entering in through "windows". It is going to be a worldwide cleanup operation, so to speak—one that shall not be completed in a single day.

The saints of God will judge and punish the world, during the day of the LORD.

153

False doctrine darkly conceals the truth

The prevailing opinion—which is largely due to the dominant narrative inculcated by so many teachers, holds: that the saints will come with Jesus Christ—but not until the time of his Second Coming, at the end of the 'Tribulation'; and, that Jesus—in one moment or, at most, in one day—is then going to destroy the armies of Antichrist in the battle of Armageddon.

There is not the least suggestion, in that narrative, that Christ and his saints shall (otherwise) be involved in executing the judgments of God, during the day of the LORD. That fictious narrative instead claims that Christ and the saints will be preoccupied (feasting)—somewhere other than earth, throughout the day of the LORD.

Not only does such teaching completely overlook the wealth of Biblical revelation concerning the coming Judgment of the world. But it completely ignores the principally important role of the saints, whom God has promised to "honor" them by committing unto them the *execution* of His Judgment, **against both men and angels, during** the day of the LORD.

The "dominant narrative", above alluded to, thus **has the effect to conceal** the true purpose and nature of the day of the LORD. Rather than to reveal *God sending* Antichrist—for the *purpose* of sealing unto eternal damnation all who will then give their allegiance to that devil: the mainstream narrative, instead, falsely gives the impression that Christ's coming is for the sole purpose to *defeat* Antichrist and, so, to deliver humanity from the *oppression* of Antichrist's malevolent rule.

According to the popular End-time fable: mankind is *not* therefore in league with the devil. Rather, mankind—the darling of God's "**un**conditional" love (as that is also dangerously supposed), will be rescued from the devil's oppression, at the very end of the 'Tribulation'.

The day of the LORD is thus transmuted—from being (as it is in truth) the Judgment of the world—into something very different. The day of the LORD is no longer, in that false narrative, the outpouring of God's "fierce anger" (Zephaniah 3:8) <u>against a generation of wicked men</u>. Instead, the day of the LORD has essentially been turned into the judgment of the Antichrist and his armies—only. Though even that punishment shall not occur until after that Christ and his saints shall have finished 'feasting'; according to the popular myth.

Throughout the past nearly thirty years, contemporary 'Christian' culture has been greatly influenced by the wildly popular *"Left Behind"* series of books and movies. Tragically, the main themes conveyed by those books and films stand in blatant contradiction to vital Biblical truths, pertaining to eternal realities involving the day of the LORD. The Bible teaches that the day of the LORD is going to be a near-extinction-level event (or, rather, series of events). Yet, the *Left Behind* series implies that the 'Tribulation' may be survived by multitudes, though not without some hardship.

Far more important, the Bible teaches that the door of God's saving mercy toward the Gentiles shall be closed, after the Rapture and throughout the day of the LORD. But the *Left Behind* series boldly denies that truth, insisting, instead, that those who miss the Rapture will be given a 'second chance' to be saved, during the 'Tribulation'. The authors and publishers of the *Left Behind* series were not the first to concoct that dangerous falsehood; though they have done more to promulgate it than did anyone else before them. That awful deception has now become a monumental 'truth' amongst many, if not in fact amongst most, churches.

The problem with rubrics (labels)

A number of key words and phrases have come to be commonly used in the manner of rubrics (labels), to represent important doctrinal truths of Scripture, i.e.: the **Trinity**; the **Rapture**; the **Tribulation**; the **Second Coming**. Each of those rubrics is supposed to embody the entirety of Biblical teaching

155

involving its respective topic. Though none of those rubrics, as such, appears in Scripture.

That latter fact alone does not invalidate the use of rubrics to suggest a certain body of teaching—but only in the case that any given rubric may be rightly related to the truth of God's Word, in the hearing and understanding of one who is taught.

The phrase, "the Second Coming", for example, cannot be found in the Bible. Nevertheless, that phrase, by itself, may suggest a whole set of ideas within an even broader context; the meaning and interpretation of which depends upon each individual's experience and beliefs. Such rubrics as just mentioned, are fraught with implications that have little if anything to do with what the Bible actually teaches. Thus, the use of rubrics, in most instances and for most people, tends to place some very substantial stumbling-block/s, so to speak, in the pathway to understanding Biblical truth.

The reality of Jesus Christ's return

The Second Coming of Christ does not occur at the Battle of Armageddon, at the end of the day of the LORD. Nor shall the solar nova initiate the day of the LORD. Rather, the day of the LORD shall begin when Jesus Christ returns—to judge the world and set up his own Kingdom on earth. *That* is the initiating Event. That is the Second Coming of Christ. Both the solar nova and the Rapture will be the result—not the cause—of Christ's return. When Jesus comes to receive his then-living saints, in the Rapture, he will not then depart the earth with his saints in tow:

> "And to you who are troubled rest with us, when the Lord Jesus shall be revealed from heaven with his mighty angels, in flaming fire taking vengeance on them that know not God, and that obey not the gospel of our Lord Jesus Christ: who shall be punished with everlasting destruction from the presence of the Lord,

and from the glory of his power; when he shall come to be glorified in his saints...."

In the passages, below, the Word of God describes the Lord Jesus Christ—in terms that very likely seem strange, even to many who profess to know him:

"The Lord is a **man of war**: the Lord is his name."

EXODUS 15:3

"Who is this King of glory? The LORD strong and mighty, the LORD **mighty in battle**."

PSALM 24:8

"Who is this that cometh from Edom, with **dyed garments** from Bozrah? this that is glorious in his apparel, travelling in the greatness of his strength?

"I that speak in righteousness, mighty to save.

"Wherefore art thou **red** in thine apparel, and thy garments like him that treadeth in the winefat? [i.e., blood-stained]

"I have trodden the winepress alone; and of the [natural] people there was none with me: for I will **tread them** in mine anger, and **trample them** in my fury; and **their blood shall be sprinkled upon my garments**, and I will stain all my raiment. For **the day of vengeance** is in mine heart, and the year of my redeemed is come. And I looked, and there was none to help; and I wondered that there was none to uphold: therefore mine own arm brought salvation unto me; and my fury, it upheld me. And I will tread down the people in mine anger, and make them drunk in my fury, and I will bring down their strength to the earth."

ISAIAH 63:1-6

"For if we sin wilfully after that we have received the knowledge of the truth, there remaineth no more sacrifice for sins, but a certain fearful looking for of judgment and fiery indignation, which shall devour the adversaries. He that despised Moses' law **died without mercy** under two or three witnesses: Of **how much sorer punishment**, suppose ye, shall he be thought worthy, who hath trodden under foot the Son of God, and hath counted the blood of the covenant, wherewith he was sanctified, an unholy thing, and hath done despite unto the Spirit of grace? For we know him that hath said, *Vengeance belongeth unto me, I will recompense, saith the Lord.* And again, *The Lord shall judge his people.* **It is a fearful thing to fall into the hands of the living God**."

HEBREWS 10:26-31

"I will therefore put you in remembrance, though ye once knew this, how that the Lord, having saved the people out of the land of Egypt, **afterward destroyed them that believed not**. And the angels which kept not their first estate, but left their own habitation, he hath reserved in everlasting chains under darkness unto the judgment of the great day.... Enoch also, the seventh from Adam, prophesied..., saying, Behold, **the Lord cometh with ten thousands of his saints, to execute judgment upon all**, and to convince all that are ungodly among them of all their ungodly deeds...."

JUDE 5-7; 14-15

Jesus is very soon returning to take full possession of, and to reign supreme over, all the earth. In the process of which—and it shall indeed be a *process*—Christ will then bestow great honor upon his faithful saints, as well as his faithful angels, by giving to them the privilege, with himself, to put down all rebellion among both men and angels and all the hosts of hell.

Of course, the Lord Jesus could by his own power obliterate the earth and all that is in it, in one instant. But he has chosen, for his own reasons, to subdue the earth—and to honor his own People, by different means. Christ, as Lord over his Creation and King over his kingdom, is teaching eternal beings, whom he has created, how to live and to serve, in the worlds he has created. Christ's return is about much more than punishing the wicked. In so doing, he is teaching principles of righteousness, not only by precept, but as well by example and experience, in his People. Throughout all of his Creation.

God waited, in the days of Noah

Most people on earth live as though they believe that—if God does exist, he is mostly indifferent to the selfishness, and debauchery, and cruelty of men, in this present world. But God's waiting does not mean he is indifferent.

More than a century before God destroyed the world by the Flood, he commanded Noah to build an ark. God's anger, even at that time, had already resolved him to abolish that world. Still, he waited for yet another hundred years or more to send that judgment upon the earth.

How white-hot must be God's anger, in this present generation? Anyone who believes in the reality of God, has not much to wonder about that question.

For more than a hundred years, Noah's neighbors were reminded every day, by the incessant pounding of hammers and raking of saws—of Noah's occasional warnings to them. One day, though, those neighbors awoke to hear nothing coming from the direction of Noah's backyard, but the peaceable sound of silence. Not realizing that that silence meant that, within a few days, they would see the sun, and feel the gentle breezes, and look upon the faces of their children, for the very last time.

Lately, there is another kind of silence in the earth, coming from the direction of America's pulpits.

That is not a good thing. But it is a sign.

159

Thief in the Night

"BUT AS THE DAYS OF NOE [NOAH] WERE, SO SHALL ALSO THE
COMING OF THE SON OF MAN BE. FOR AS IN THE DAYS THAT
WERE BEFORE THE FLOOD THEY WERE EATING AND DRINKING,
MARRYING AND GIVING IN MARRIAGE, UNTIL THE DAY THAT
NOE ENTERED INTO THE ARK, AND KNEW NOT UNTIL THE FLOOD
CAME, AND TOOK THEM ALL AWAY; SO SHALL ALSO THE
COMING OF THE SON OF MAN BE."

MATTHEW 24:38-39

There are many—even amongst professing Christians—
who adamantly deny it is possible to know, with any degree of
certainty, the time of Christ's return. Such are evidently blind
to the fact that a *large percentage* of the Bible consists of
divinely-given prophecies pertaining to the Second Coming of
Christ. Such persons are blind as well to the purpose of God's
Word. Prophetic Scripture was not given to mystify or
confound God's obedient children but, rather, to enlighten
them:

> "For yourselves know perfectly that the day of the Lord
> so cometh as a thief in the night.... But **ye, brethren,
> are not in darkness, that that day should overtake you
> as a thief**. Ye are all the children of light, and the
> children of the day: we are not of the night, nor of
> darkness."
>
> 1 THESSALONIANS 5:2-5

"As a thief in the night...", means that Christ's return shall
take the unbelieving world (including multitudes of careless
professors of religion) by complete surprise—for a reason that
itself is surprising to most. Some believe that conditions in the
world are not yet bad enough to warrant Christ's return. Still
others believe the Church must first subdue the world and
prepare the Kingdom, which Christ shall then return to rule.

160

But the truth is nothing like what is suggested by either of those myths. Jesus himself said that, in the days leading up to his return, *it will be business as usual* in the world. That was the startling warning Jesus declared when he likened the time of his return, to the days of Noah. Christ's return to earth is going to take the unbelieving world by surprise—*"as a thief in the night"*: because, his return will occur suddenly, on a day not unlike every other day.

> "The Lord is not slack concerning his promise, as some men count slackness.... But the day of the Lord will come as a thief in the night...".
>
> 2 PETER 3:9-10

> "Likewise also as it was in the days of Lot; they did eat, they drank, they bought, they sold, they planted, they builded; but **the same day that Lot went out** of Sodom it rained fire and brimstone from heaven, and destroyed them all. **Even thus shall it be in the day when the Son of man is revealed.**"
>
> LUKE 17:28-30

Yet, to say it will be "business as usual", or "a day not unlike every other day", when Christ will come to judge the world, does not of course imply that the world, at that time, shall be tranquil. It simply means that no dramatic signs will appear to alert an unbelieving world, when Judgment Day arrives.

The Bible unambiguously portrays the psycho-social condition of human society, in the days of Jesus's return, in terms of: pervasive self-centeredness and pleasure-seeking—overshadowed by unrelenting despair, in the face of widespread lawlessness and societal breakdown:

> "This know also, that in the last days perilous times shall come. For men shall be **lovers of their own selves**, covetous, boasters, proud, blasphemers, disobedient to parents, unthankful, unholy, without natural affection, trucebreakers, false accusers,

161

incontinent, fierce, **despisers of those that are good**, traitors, heady, highminded, **lovers of pleasures** more than lovers of God…."

2 TIMOTHY 3:1-4

It would be difficult to conceive a more apt description of the world as it is today. Which appears to closely resemble what the world was like in the days preceding the Flood—as seen in the following passages of Scripture:

> "And GOD saw that <u>the wickedness of man was great</u> in the earth, and that **every imagination of the thoughts of his heart was only evil continually**. And it repented the LORD that he had made man on the earth, and it grieved him at his heart. And the LORD said, I will destroy man whom I have created from the face of the earth; both man, and beast, and the creeping thing, and the fowls of the air; for it repenteth me that I have made them. But <u>Noah</u> found grace in the eyes of the LORD….
>
> "The earth also was corrupt before God, and **the earth was filled with violence**. And God looked upon the earth, and, behold, it was corrupt; for <u>all flesh had corrupted his way</u> upon the earth. And God said unto <u>Noah</u>, *The end of all flesh is come before me; for <u>the earth is filled with violence through them</u>; and, behold, I will destroy them with the earth.*"

GENESIS 6:5-8; 11-13

Civilization, so-called, is not progressing—regardless of what technological marvels man has produced. It has long been regressing, decaying, rotting, until that, now, human society appears to be teetering on the brink of total collapse.

The most fundamental realities related to biological differences between male and female, for one important example, are no longer valid considerations in the dystopian vision of global leaders' intended New World Order. The rapid

162

development of computer technologies, including Artificial Intelligence (AI); coupled with equally impressive developments in the realms of Biochemistry and Materials science: are tools that world leaders and their myriad followers are determined to use in creating a new Humanity consisting entirely of transhumans: an integration of human biology and electro-mechanical technologies: the fusion of man and machine. This is not a joke.

The rapidly emerging transexual revolution is not some quirky fad. It is a key strategy that is being used to destroy Society's concept of what it means to be 'human'. It strikes deep at the heart of the meaning of human identity and personhood. It moreover constitutes a paradigm shift with respect to how (the world is being programmed) to view, in a radical new way, the supposed mutability and 'design-ability' of biological organisms—of which human beings are but one example. *Science as religion* has deceived its priests and prophets to believe that human life is fully within their purview and power not only to control but virtually to re-create.

There are no longer *human persons*. There are only biological *entities*, subjects for scientific experimentation. The horrific proofs of which, heinous beliefs, are more and more coming to light. Contemporary culture, especially in developed nations, is not merely decadent, it is abominable, abhorrent.

(Was there no such 'Science' in the pre-Flood world— which, somehow, managed to breed a veritable *race* of giants? See, Genesis chapter 6.)

The present madness just alluded to notwithstanding; the perpetual-motion machinery of life, so to speak, must go on. There are bills to pay, and one must eat to live. Nations, and cultures, and communities; governments, and businesses, and schools; wars, and sports, and jobs; all the varied interests and activities that incessantly animate nearly 8 billion souls: the restless motions of human life will continue on in the general 'current' of things—right up until the very day when Jesus appears. Without any other warning than what has been given in Scripture.

Jesus is very soon going to come, on one such, indistinct day. It was not raining when Noah and his family entered the Ark. The Bible even seems to suggest that, in the time before the Flood, 'rain' as such did never fall from the sky, but *"there went up a mist from the earth, and watered the whole face of the ground"* (Genesis 2:4-6). Aside from Noah's own family, no one was expecting "the windows of heaven" to break open, or the "fountains of the great deep" to burst forth, on that or any other day.

> "[A]nd [they] **knew not until** the flood came, and took them all away; *so shall also the coming of the Son of man be.*"
>
> JESUS CHRIST (QUOTED IN MATTHEW 24:39)

There are multitudes who believe they have plenty of time to get right with God, because, they expect to see some kind of 'sign' before Jesus comes to judge the world; or, because they believe that such-and-such conditions must prevail in the world, before Christ shall appear. Many of them, perhaps, suppose themselves to be Christians (at least, they seem to believe in the reality of a coming judgment). But they are indifferent to God's Word, apathetic concerning the truth, insensitive to God's Spirit. Tragically, such persons will still be waiting for a sign—when the day of the LORD suddenly breaks upon them without warning.

The monotony of daily struggling for survival tends to wrest the souls of men and women towards enslavement to a base existence. One would think that oppressed souls would leap for joy, at the opportunity to receive the free gift of God's salvation, of his own Spirit and life. But humans are by nature children of the world and, so, love the world with its pleasures and distractions. The incalculable void; that infinite place in each and every soul; *that part of us that was created to be God's dwelling place*: that vast eternal emptiness requires to be filled with—

164

something. Entertainments and pleasures are easily obtained to stimulate one's senses. But how can the knowledge of God satisfy? How can belief in God supply one's daily needs? What good is it to hope in some unseen and unknowable afterlife, meanwhile trying to cope with the necessities and hardships of this present life…?

> "Multitudes, multitudes in the valley of decision: for the day of the LORD is near in the valley of decision."
> JOEL 3:14

Multitudes truly are in the valley of decision. But the day of the LORD is near—in the valley of decision. When the Sixth Seal events suddenly occur without warning, it will then be too late to decide. While it is true that everyone has the right to decide—whether for, or against, Christ; yet, no one has the right *not* to decide. A decision to postpone dealing with Christ *on his terms*, is the same as a decision to reject him. And all those who *on that day* will still be found in the valley of decision, shall then discover their eternal destiny has finally been decided for them.

Whether you or I may live to see all these things come to pass in our lifetime (which we very likely shall), the question still remains:

Are you ready for Eternity?

---

www.ingramcontent.com/pod-product-compliance
Lightning Source LLC
LaVergne TN
LVHW051238080426
835513LV00016B/1647